THE DOSE MAKES
THE POISON

THE DOSE MAKES THE POISON

A Plain-Language Guide to Toxicology

THIRD EDITION

PATRICIA FRANK, PH.D.
M. ALICE OTTOBONI, PH.D.

WILEY

A JOHN WILEY & SONS, INC., PUBLICATION

Published by John Wiley & Sons, Inc., Hoboken, New Jersey
Published simultaneously in Canada

For general information on our other products and services or for technical support, please contact our Customer Care Department within the United States at 877-762-2974, outside the United States at 317-572-3993 or fax 317-572-4002.

Wiley also publishes its books in a variety of electronic formats. Some content that appears in print may not be available in electronic formats. For more information about Wiley products, visit our web site at www.wiley.com.

Library of Congress Cataloging-in-Publication Data:
Frank, Patricia.
 The dose makes the poison : a plain-language guide to toxicology / Patricia Frank, M. Alice Ottoboni. – 3rd ed.
  p. cm.
 Includes bibliographical references and index.
 Prev. ed., published in 1991, entered under M. Alice Ottoboni.
 ISBN 978-0-470-38112-0 (pbk.)
1. Toxicology–Popular works.   I. Ottoboni, M. Alice.   II. Ottoboni, M. Alice. Dose makes the poison.   III. Title.
  RA1213.O88   2011
  615.9–dc22

                                                                    2010023287

oBook ISBN: 978-0-470-91844-9
ePDF ISBN: 978-0-470-91843-2
ePub ISBN: 978-0-470-92273-6

10 9 8 7 6 5 4 3 2

CONTENTS

INTRODUCTION TO THE THIRD EDITION

When I was asked to revise this book for the third edition, I wanted to increase its scope from dealing primarily with environmental chemicals to dealing with all types of chemicals that we confront every day, not only environmental chemicals, but also drugs, food additives, vitamins, and others. I have included most of the preface and introduction to the second edition so that you can see Alice Ottoboni's original intention and also her insights into toxicology that were presented there.

I think that broadening the scope is important because each day I read several newspapers, newsmagazines, the Internet news, various blogs, and so on, and each day I see articles about all the "poisons" in our world. Some of the articles point out which pesticides are found as residues on peaches, how much lead is in Barbie's shoes, what drugs are found in our drinking water, and other facts or nonfacts. Other articles have a viewpoint to share and it seems their raison d'être is to scare the public. As I read them, I can imagine how disturbed a nontoxicologist would be to see all the headlines and not know what to do about these frightening things. The days are long over when we can be comfortable that our food supply, our water supply, our drugs, vitamins, and cosmetics are as safe as we would like. So it is important to increase our understanding of all types of chemicals in our ever more complex world.

Our society has made great strides in controlling the unbridled spread of various chemicals into the environment. From the removal of arsenic from cosmetics and drugs in the early 1900s, the establishment of the Environmental Protection Agency in 1970 in order to control the use of pesticides and set standards for air and water quality, up to the empowerment of a modern Food and Drug Agency that monitors our medicines and medical devices, we have taken steps to ensure the safety of many products.

Some of us were lulled into complacency that the world was becoming a safer place. That balloon burst in the 1980s when there were some terrible industrial accidents such as at a Bhopal, India, factory as well as consumer product tampering, all of which led to sickness and deaths of innocent bystanders. Today, with the advent of serious industrialization of nonwestern economies such as China and India, the integrity of our food supply is once again in question and the safety of various consumer products and medicines may be at risk.

Lest we be lulled into a sense of complacency over environmental safety, toxic waste spills still occur with serious impact on both humans and wildlife. For example, two of the headlines in 2009 concerned an ammonia leak that killed someone who drove into the gas cloud and the hospitalization of a worker exposed to a large spill of aniline. As I am finishing this book, I am watching with horror the beginnings of what may be the worst oil spill in U.S. history, namely the sinking of an offshore oil rig in the Gulf of Mexico and the consequences to our environment and economy from millions of gallons of oil washing ashore along the coastal states. The long-term consequences of wildlife and human exposure to the inhalation of oil fumes and the ingestion of oil residues in our food and water are certain to be discussed for the next decades. This is all the more reason for the public to understand how to assess risk in order to determine when to be curious, when to be nervous, and when to be truly scared.

Because I have spent most of my career working in the area of pharmaceutical development, many friends ask me questions about their medications. My response always starts with "What did your doctor say?" And the answer is usually either "I didn't ask" or "He/she didn't know." Most people are surprised to find that there are side effects from drugs. I tell them that *all* chemicals have side effects. It is only where on the scale of toxicity a chemical falls. My very simple, on-a-cocktail-napkin scale is:

So where does your medicine, food, water, pesticide fall on this scale?

I hope by the end of this book that you will be able to use the information presented in order to be a more informed consumer of food, water, and medicinal products and to be able to establish your own risk-benefit scenarios for many aspects of your life. In the last chapter of the book we take an in-depth look at how to assess risk while in Chapter 8 we present a variety of case histories that explicate some of the issues surrounding the use of chemicals in our industrial society. Some of these examples are a bit old, but they still have relevance as cautionary tales.

As always with an undertaking such as this, there are many people to thank. First, Alice Ottoboni, whose first two editions laid all the groundwork for this edition, has been a great source of help and encouragement. I also appreciate the input of our editor, Jonathan Rose, and his staff and the critical eye of Barbara Flynn-Waller. Of course, I am grateful to my husband, Jerry, for his thoughtful comments and consistent encouragement, not only for this book but during my whole career.

PATRICIA FRANK, 2010

PREFACE TO THE SECOND EDITION

The natural laws that direct the orderliness of our world are part of our everyday lives. People know that water always runs downhill, that apples always fall to the ground when their stems break, and that the sun always traverses the sky from east to west. Natural laws are immutable, constant, and predictable. So it is with the laws that govern the behavior of chemicals, natural or synthetic. The toxic effects of a given chemical depend on dose (how much), frequency of exposure (how often), and the route by which the chemical enters the body. It always has been thus, and there is no reason to believe it will ever be otherwise. Yet some people find it difficult to believe that chemicals follow any rules at all.

The laws that govern the toxicity of chemicals do not readily manifest themselves in our daily routines, with the result that we have little knowledge or awareness of them. Thus, at the end of World War II, when the rapidly developing petrochemical industry presented us with a host of new synthetic chemicals, whose names we could not pronounce and with which we were unfamiliar, a certain segment of our population became suspicious; anything man-made was viewed with mistrust. Then, in the early 1960s, when the public media began its intense and continuing focus on private worries and concerns that synthetic chemicals were causing great damage to wildlife, the environment, and even us humans, many people became frightened.

Fear of many synthetic chemicals has not abated, despite a lack of objective evidence that they have been detrimental to the public health. Americans are living longer and are healthier than ever before in our history. Nevertheless, a significant segment of our population still believes that many synthetic chemicals are harming them and threatening them with cancer. Adults who remain resistant to chemophobic fears for their own health are challenged to come into the fold with stories of dire consequences for their children. Recently, claims of immune system damage and loss of reproductive capability from exposure to synthetic chemicals have added to the burden of fear.

This book was written with a firm conviction that fear of certain chemicals, or more specifically fear of certain synthetic chemicals, is the product of a lack of understanding of the naturals laws that govern toxicity and, as a corollary, a conviction that knowledge of what makes a chemical harmful or harmless can help dispel unreasoning fear and aid in our dealing more effectively with some of the real problems related to chemical exposures.

I have found, from my many years of working for and with the public, that most people are intelligent and perceptive individuals who want scientific facts relating to subjects that are vital to their health and well-being. Even without scientific education, they are completely capable of understanding such facts. This book is for them. Its purpose is to provide facts about the toxicity of chemicals and to help people to cope with news and media reporting, preserve their sanity in the face of poison paranoia, and make informed judgments about chemicals in the environment.

The public's fear of chemicals, combined with increasing recognition within government and industry that people must be protected from harmful exposures to chemicals, has resulted in a dramatic increase during past decades in the number of laws regulating environmental chemicals. There has been little substantive change in these laws, outlined in the section "Regulation of Toxic Chemicals," since the publication of the first edition of this book. Through the years, there have been numerous challenges to the Delaney Clause, with groups fearful of human exposures to synthetic chemicals lobbying to expand its coverage, and groups concerned that the Delaney Clause excludes scientific judgment in its implementation lobbying to eliminate it. The matter was put to rest for the time being with the passage, in August 1996, of the Food Quality Protection Act, which supplants the Delaney Clause. This change, while of significance to the lobbyists, both for and against the Delaney Clause, will probably have little or no impact on public health. However, the debate has begun anew about the benefits and detriments of the change.

I am grateful to the many friends and associates with whom I have discussed this second edition for their very valuable comments and criticisms. I owe a special debt of gratitude to my husband, Fred. His sharing of his knowledge of public, occupational, and environmental health has been of tremendous benefit to me not only in the preparation of this second edition but in all of my professional activities.

INTRODUCTION TO THE SECOND EDITION

Many years of service as a public health toxicologist for the California Department of Public Health (now the Department of Health Services) made it disturbingly clear to me that an inordinate fear of chemicals was the rule rather than the exception among the general public. During the same years, participation in training programs designed to teach people how to work safely with the chemicals they contacted in their occupations taught me that people with no science background were not only capable of understanding the basic principles of toxicology but that they could also apply what they learned to work safely and comfortably with some very dangerous chemicals. This book was born of these two observations.

There is a general lack of public understanding about what makes chemicals toxic, and about the word that has become a synonym for *toxic*. That word, now a part of our everyday vocabulary, is *poison*. Headlines tell us about the poisons in our food, poisons in our water, poisons in our air; poisons everywhere! People who use the word most freely appear to have the least concept of what poison means. The indiscriminate use of the word has brought us into an era of what might be termed *poison paranoia*.

Whenever some misfortune occurs for which we have no ready explanation—an illness, a mischance of nature, a declining wildlife species—we look to blame some chemical. This propensity is aptly illustrated by the mystery of the double-yolked eggs, reported in the Consumers Cooperative of Berkeley newspaper, the *Co-op News*, July 16, 1979: "Science is beautiful, but it can sometimes spoil a good news story." The story went on to tell that a Co-op member was recently amazed when she found NINE double eggs out of a dozen box. I shared her astonishment, convinced that either the odds against this marvelous happening were billions to one or that some horrible chemical additive fed to a chicken had caused it and that some serious muckraking was needed down at the chicken ranch to protect embattled consumers by eliminating this poison from their diet.

The Co-op home economist checked with the supplier of the eggs and received a reply that took all of the mystery out of the event by placing it squarely in the dull world of young chickens and egg sorting, where neither chemicals nor miraculous odds were at issue.

Young chickens are apt to pop more eggs with two yolks, but it becomes more uncommon as they reach maturity. The reason nine eggs could wind up in the same box is because double eggs are oversized, so they get set aside by the egg sorter because they won't fit in the egg container. However, there are some borderline ones which the sorter selects from those set aside and allows to pass through. This is why so many were in one box.

Fortunately, in the case of the double-yolked eggs, further facts were sought and the real reason for the apparent anomaly was discovered, thereby avoiding another scare headline. Unfortunately, such dedication in pursuit of truth is often the exception rather than the rule.

There are two diametrically opposed dangers in news media toxicology and its offspring, poison paranoia. One is the cry-wolf syndrome. When an alarm is sounded frequently and without regard to degree of emergency, the alarm becomes meaningless and, therefore, is not effective when a true emergency exists. It is well known that to call everything bad, in effect, is to call nothing bad. If safe and sane use of chemicals in our homes, work, and recreation places is to be furthered, there must be understanding, cooperation, and support on the part of the public. A public blasé about harmful effects of chemicals is a public disinterested in making any changes in use practices relating to chemicals. Such a public attitude would be tragic.

The second danger is that a certain fraction of our population will become victims of a helpless, hopeless fear and terror that chemicals from which they cannot escape—chemicals in their food, their water, their air—are destroying their health, shortening their lives, or dooming them to cancer. Such a fear is a form of stress that can be just as damaging as the chemicals that are feared, and in some cases even more so. Stress can produce vague feelings of illness, such as nausea, headache, weakness, and malaise, as well as actual physical illness. The medical profession now generally accepts the premise that stress can exert a profound influence on the course of many illnesses and appears to trigger or worsen some diseases, such as high blood pressure and Crohn's disease (a type of colitis).

Poison paranoia already is taking a toll in the mental health and well-being of some people. This conclusion is based on the many thousands of calls, letters, and visits that I have received from people concerned about the health effects of chemicals in their environments. The gamut of their concern extended from calm interest to outright panic. In a few cases, the cause for apprehension was valid because, through some accident, misuse, or lack of knowledge, there had been an actual or potential exposure to a harmful level of some chemical. However, in the majority of cases, the fears or concerns were ill-defined and prompted, in the main, by the most recent scare headline. Among the latter, there were a few people who refused to accept any information that did not support their conviction that they were suffering from some sort of chemical poisoning. People who fall victim to an unreasonable fear of chemicals are literally frightened sick. Frightened people truly suffer. They are victims of distorted information and lack of knowledge.

The great majority of people are seriously concerned about the many chemicals reported to be harming them and the environment, but they do not have a pathologic fear about the effects of chemicals on their health. For the most part, they do not know what to do about the situation, other than to modify their lifestyles to the extent possible. They can live without smoking, but they cannot live without breathing.

This book is not intended as a condemnation of, or an apology for, synthetic chemicals; rather its aim is to present an objective discussion of what makes chemicals harmful or harmless. I feel compelled to make this point so that the reader will understand that I hold no brief for or against synthetic chemicals; they are facts of life with which we must deal. I have learned from many years of contact with people of all viewpoints regarding the risks posed by chemicals that objectivity often invites scorn from both extremes of view. Thus, both pro- and anti-chemical extremists may take exception to all or parts of this book because it is not directed toward reinforcement of their respective "what's-the-fuss" and "ain't-it-awful" views. This book is not written for people of extreme persuasions but rather for people who want a real understanding of the significance of their many chemical exposures. Only people with open minds are tolerant of concepts that are new to them or in conflict with their beliefs.

The comfort provided by knowledge was vividly brought home to me many years ago by a young woman who called for information about a chemical. After a rather lengthy conversation, she said, "I feel so sorry for you. You know so much about all the harmful effects of the chemicals that surround us that you must really worry all the time."

I was surprised by her statement, because such a thought had never occurred to me. I assured her that, on the contrary, the very fact that I do know what makes chemicals harmful frees me from worry. All chemicals follow the same rules: the laws of nature. By knowing the rules, I have a perspective that protects me from needless worry and unreasoning fear. My hope is that this book will give you the same perspective.

M. ALICE OTTOBONI, 1997

CHAPTER 1

WHAT ARE CHEMICALS?

The word *chemical* has become a dirty word in our modern American vocabulary. Our public media provide us daily with advice or warnings about the presence of chemicals in our food, air, and water and the harm they are doing to us and the world we live in. As a result, the word *chemical* conjures up visions of damage, debility, disease, and death in the minds of many people. In order to understand the threats posed by chemicals—a prerequisite to wisely protecting ourselves and our environment from their adverse effects—we must clarify or reform our concept of *chemical*.

ATOMS AND MOLECULES

All matter is composed of chemical elements. An individual unit of an element is called an atom. Atoms are the basic building blocks for all substances. Approximately 90 different kinds of stable elements are found in nature. Examples of elements are hydrogen, oxygen, carbon, nitrogen, gold, and silver. A complete listing of all of the elements, including those that are unstable (radioactive), can be found in any good dictionary. The periodic table gives detailed information about all the elements and the relationships among them. A multicolored diagram of the periodic table and an explanation of how this table is constructed can be found at the Los Alamos National Laboratory Web site (http.//periodic.lanl.gov) or the University of Sheffield Web site (www.webelements.com). Appendix A describes the concept of Avogadro's number and molecular weights for those who might be interested.

When two or more atoms (usually of different elements) are linked together by chemical bonding, they form units called molecules. A substance composed of molecules all of the same kind is called a compound. Water, salt, and sugar are examples of compounds. The number of different kinds of molecules that can be formed by the combination of from two to many thousands of atoms, from more than 90 different elements, is astronomical. Figure 1-1 shows the structures of a very simple and a very complex

The Dose Makes the Poison: A Plain Language Guide to Toxicology, Third Edition.
By Patricia Frank and M. Alice Ottoboni

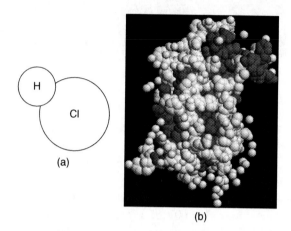

FIGURE 1-1 (a) Hydrochloric acid, a simple compound; (b) growth hormone, a complex compound. [Part (b) from Wikimedia open source, http://commons.wikimedia. org.]

molecule. All substances are composed of chemical and physical combinations of atoms (elements) and molecules (compounds). Thus, everything in our physical world is chemical—the food we eat, the water we drink, the clothes we wear, the medicines we take, the cosmetics we use, the plants in our garden, our furniture, our homes, our automobiles, and even ourselves. Our entire physical world is composed of chemicals.

NATURAL CHEMICALS

The total number of chemical compounds in our universe that occur naturally will never be known exactly, but, from the millions that have been identified thus far, we know that the total number is huge. Natural chemicals may be organic (i.e., containing carbon) or inorganic. Our inanimate world is an inorganic world. It is composed of a great number of mineral substances in which all of the elements, except for a few radioactive elements that have been created by nuclear scientists, are represented.

Our living world is composed primarily of organic compounds, the diversity of which is tremendously greater than that in our inorganic world. The number of natural organic compounds that has been identified thus far, although very large, is probably negligible compared to the number of those yet unidentified. Many of these as-yet-unidentified organic chemicals—components of the trees, shrubs, and other plants of the rain forests—could well be of great value to medical and pharmaceutical sciences.

One small segment of our organic world, food plants and animals, provide us with the nutrients that we use to build and repair our bodies. However, the plants and animals we use for food contain many more natural chemicals than just the nutrients we require. Since it is impossible to separate nutrients from non-nutrients in our foods, we depend on our bodies to do this work for us. There are many kinds and quantities of nonnutrients in our foods, particularly our plant foods. The animals we use for food have already done the job for us of selecting nutrients and eliminating most of the nonnutrients from plants.

Among the natural chemicals that we eat, many can cause adverse effects if consumed in excess. In fact, there is probably no food that does not contain some potentially harmful natural chemical. This fact is the basis for an annual project of the American Council on Science and Health (ACSH).* Every fall the ACSH publishes a typical Thanksgiving menu accompanied by identification of the naturally occurring toxic or carcinogenic chemicals present in each food on it (found at www.acsh.org). For example, taken from the 2009 menu are heterocyclic amines, acrylamide, benzo(a)pyrene, ethyl carbamate, dihydrazines, d-limonene, safrole, and quercetin glycosides—and this just from the turkey with stuffing! If you are keeping to a vegetarian diet, then the 2009 menu shows salad may contain aniline, caffeic acid, benz-aldehyde, hydrogen peroxide, quercetin glycosides, and psoralens.

An interesting method for ranking the potential health effects from exposure to such toxicants that occur naturally in foods was developed by Bruce Ames and his colleagues at the University of California, Berkeley. Dr. Ames has written numerous articles for both scientific and popular publications reviewing the subject of naturally occurring toxicants and their carcinogenic hazards. Rankings are based on data from the scientific literature as well as from Dr. Ames's own laboratory, using accepted methods of risk assessment. These rankings are one approach to the evaluation of relative health risks posed by suspected carcinogens, both natural and synthetic.

SYNTHETIC CHEMICALS

Humans, in their ingenuity, have been able to take the basic building blocks of which all matter is composed and link them together in new combinations to produce compounds not found in nature. Thus, we have a host of synthetic substances, primarily organic, available to us, which we put to a seemingly endless variety of uses—pharmaceuticals, pesticides, and polymers of all sorts, including the common household plastics with which we are so familiar.

* A list of abbreviations can be found at the end of the book.

The term *organic* has been extensively used by the health-food industry to mean one thing and used by chemists to mean another; as a result, the term is generally misunderstood by the public. *Organic* has come to mean something (usually food) that is naturally occurring or produced without the use of pesticides or other synthetic chemicals, such as hormones. Scientifically, organic chemicals are simply chemicals composed primarily of the element carbon, independent of whether they are natural or synthetic. It comes as a shock to many people that almost all synthetic chemicals, including pesticides, are organic chemicals. The term *organic* was coined long before the birth of modern chemistry.

Early scientists who studied the composition of matter recognized that substances produced by living organisms were different from all other chemicals then known to humans. They called the former *organic* (derived from organisms) as opposed to the latter, which they classified as *inorganic*. Early in the nineteenth century, scientists discovered that the element carbon was present in all organic compounds; hence carbon chemistry became synonymous with organic chemistry.

The great complexity of carbon chemistry, relative to inorganic chemistry, the large size and complicated structures of many organic compounds, their great number and variety, combined with the fact that organic chemicals were found only in living organisms or products of living organisms led the early-day chemists to endow organic chemicals with mystical properties. They considered that the laws that governed the behavior of inorganic chemicals did not apply to organic chemicals; humans could synthesize— that is, manufacture—compounds such as nitrous oxide and hydrochloric acid but were incapable of synthesizing organic compounds in the laboratory at that time.

The special properties of organic chemicals were attributed to the action of a supernatural force, the "vital force," as distinct from the crude and vulgar forces that governed inorganic chemicals. Jöns Berzelius, a noted chemist of the early nineteenth century, wrote that the vital force was unrelated to inorganic elements and determined none of their characteristic properties. Berzelius considered that the vital force was a mysterious property beyond comprehension.

The birth of synthetic organic chemistry occurred at about the time of Berzelius's writing, with the first laboratory synthesis of an organic chemical, using basic chemicals as starting materials. The first synthetic organic chemical was oxalic acid, made by the German chemist Friedrich Wohler. A short time later in 1824, Wohler also synthesized urea. After this accomplishment, Wohler wrote to Berzelius to tell him that he had prepared urea, a chemical found in the urine of animals, "without requiring a kidney or animal, either man or dog."

The notion that organic and inorganic chemicals were qualitatively different persisted for decades after the revolutionary demonstration that humans could, indeed, synthesize organic chemicals. The science of chemistry was greatly retarded until the chemical properties of carbon and its place in the periodic table were more fully understood. The great numbers of synthetic organic chemicals that have been created since the end of World War II were not of much public interest until the publication of Rachel Carson's book *Silent Spring* in 1962. This book stimulated great interest in the effects of pesticides on environmental and public health and brought to public attention the proliferation of chemicals.

The number and variety of synthetic organic chemicals are truly amazing. In 1978, the American Chemical Society's registry of chemicals listed over 4 million organic and inorganic chemicals; of this number, more than 95 percent were organic. Of all the known organic chemicals, perhaps half are naturally occurring chemicals that have been synthesized in the laboratory or isolated from natural sources. Between 1965 and 1983, 6 million additional chemicals had been produced, and the rate of synthesis has only increased since then.

For the average person, what is the significance of the existence of these millions of chemicals? Among those that are not naturally occurring, a great many exist only in small quantities in vials on chemists' benches or in chemical storerooms. They have not been found to have any practical use or function, and so they have not been developed commercially—yet. However, with the advent of high-throughput screening techniques where robotics speed up the screening process, many thousands of chemicals can be rapidly analyzed for their ability to bind to various animal and human chemical receptors. Out of this screening, chemicals that were once thought to have no value are being identified as potential medicines and pesticides and for other human uses.

The toxicity of synthetic chemicals—that is, the degree to which they are poisonous—covers the entire range from essentially nontoxic to extremely toxic. This is also true of inorganic compounds (think water and arsenic). Some synthetic chemicals, such as artificial sweeteners, are edible, whereas others, such as chemical warfare agents, are lethal in extremely small amounts. Regardless of the degree of toxicity, the principles of toxicology apply equally to all chemicals, whether synthetic or natural, organic or inorganic.

The number of chemicals that actually enter homes is not known, but a survey of the wide variety of products found in the home setting—such as cleansers, polishes, drugs, cosmetics, prepared foods, pesticides and other garden chemicals, automotive products, and hobby products—suggests that it is quite large. Despite the wide variety of products, many contain the same basic chemicals. Thus, the actual number of individual chemicals that the average person comes in contact with in home products is probably much closer to several thousand rather than several million. The majority of chemicals that enter homes are not harmful when used properly, but some are treated with a more cavalier attitude than is warranted, as witnessed by the numerous accidental poisonings that occur in children.

The people in contact with the widest variety of potentially dangerous chemicals are those in businesses or professions that use chemicals in some process or procedure and those who work in industries that synthesize, manufacture, formulate, or use chemicals to make other products. Few of these chemicals find their way into a home setting.

CHEMICAL CATEGORIES

We categorize chemicals in many different ways, the broadest of which is whether they are natural—produced by a living process—or synthetic—made by humans. Other ways we classify chemicals are by the use we make of them (foods, drugs, pesticides, etc.), how they are physically organized (solid, liquid, gas), what kind of animal they are (fish, reptiles, birds, mammals, etc.), whether they are organic or inorganic (animal, vegetable, or mineral), and so forth. Plant and animal probably were two of the earliest categories recognized by humans. Plants stayed put, whereas animals usually moved about freely. Based on this classification, corals were considered plants for many years until their animal nature was discovered.

A scheme of classification by the use we make of a chemical or product is essential for government regulation of such items as foods, drugs, cosmetics, pesticides, industrial chemicals, and medical devices. If a substance is claimed to be a food, it is governed by the food laws. If the exact same substance is packaged and labeled a drug, it is governed by the drug laws, not by the food laws. The laws that pertain depend on what use the manufacturer or seller specifies for the product. For example, hydrochloric acid is regulated as a household product when it is present in cleaning compounds, as a drug when it is used to treat people with low gastric acidity, as a hazardous industrial chemical when it is used in electroplating, and as a antibacterial adjuvant when it is used to enhance the germicidal activity of chlorine in swimming pools. Hydrochloric acid is natural when produced by the stomach and synthetic when made in the laboratory. Interestingly, all things tobacco are regulated by the Bureau of Alcohol, Tobacco, Firearms and Explosives (ATF), but since 2009 the Food and Drug Administration (FDA) is monitoring the advertising and content of cigarettes, emphasizing the toxic nature of cigarette ingredients and smoke. (The ATF was originally part of the Department of the Treasury and was primarily concerned with collecting revenue generated by taxes on the items it regulated. ATF still is involved in investigating the smuggling of cigarettes.)

Another example is boric acid, which occurs naturally as the mineral sassolite but also can be synthesized in the laboratory. It is regulated as a household product when used in laundry detergents, as a drug when sold as an antiseptic eyewash, as an insecticide when used to kill roaches, as an herbicide when applied to kill weeds, and as a flame retardant when used to fireproof fabrics. Many chemicals, such as hydrochloric acid and boric acid, fall into both drug and pesticide categories. Coumarin compounds, such as warfarin, are not only excellent rodenticides but are also valuable anticoagulant drugs that are used to prevent blood clots. Dichloro diphenyl dichloroethane (DDD), a close relative of dichloro diphenyl trichloroethane (DDT)—the infamous pesticide now banned in the United States—and itself an insecticide, was once used therapeutically to treat certain forms of adrenal cancer.

The important lesson to be learned from these examples should be apparent: The physical, chemical, and toxicologic properties of any chemical are totally independent of the category in which it is placed. The toxicity of boric acid is exactly the same when it is used as a drug as it is when it is used as a pesticide.

Although people are concerned about the products and effluents from the chemical industry, the class of man-made chemicals that is almost universally of concern is the category known as pesticides. Pesticides are substances, natural or synthetic, that are used to kill a plant, animal, insect, or other organism that has been determined to be undesirable for some economic, medical, or esthetic reason. Included in the pesticide category are insecticides, fungicides, herbicides, rodenticides, germicides, and a whole host of other "-cides."

Countless chemicals are as toxic or more toxic than many of the pesticides, but the focus of fear centers on this group. Why? One reason is the tremendous amount of publicity given to reports of damage from the presence of pesticides in our environment and even in our own bodies. Another reason is that pesticides are used to kill living things and thus are labeled as poisons in the public mind. The concept of poison is considered by many people to be an all-or-none phenomenon: A chemical is either a poison or it is not, with no shades of gray in between. Nothing could be further from the truth. Such simplistic reasoning is counterproductive to an understanding of how and why chemicals cause harm. It also points up the fallacy of assigning blanket judgments of safety or harm to categories of chemicals.

CHEMICALS: "GOOD" AND "BAD"

A common misconception that must be overcome before an understanding of toxicity can be achieved is that chemicals made by nature are good and those made by humans are bad. Actually, toxicologists recognize that Mother Nature is far more ingenious than humans could ever be in devising toxic chemicals; it is also much more prolific. Of all the chemicals, the number of natural ones far exceeds the number made by humans. In addition, there are tens to hundreds of thousands of plants that botanists have not yet identified, much less characterized chemically. The voluminous literature on the toxic properties of naturally occurring chemicals that have been identified in food and nonfood plants, animals, and microorganisms supports an estimate that the fraction of natural chemicals that are toxic is at least as great as the fraction of synthetic chemicals that are toxic.

Some of the most toxic chemicals are produced by living organisms. A good example is botulin, the toxin produced by *Clostridium botulinum* organisms. One milligram (mg) of botulin (128 thousandth of an ounce) is capable of killing 20 million mice. It is estimated that the average oral lethal dose of botulin for an adult human is about 1 nanogram (ng) with one tablespoon containing enough toxin to kill over 3 *billion* people. Botulin toxin is available commercially as the active ingredient in prescription wrinkle injections (Botox® and others), and other medical uses have been found for it, including treating overactive bladder in children and treating muscle spasms. Studies have even been conducted for its efficacy in treating Parkinson's disease. So although botulin is a very toxic compound, it has beneficial uses for humans when used correctly.

Since we will be using a variety of measurements in the metric system, Table 1-1 introduces these units. When we refer to the concentration of a chemical in the air or solution, we use the units presented in Table 1-2.

Toxic chemicals of natural origin, such as those produced by algae and other microorganisms, snakes and other venomous animals, and plants, constitute a common threat to wild and domestic species. Actually, natural and synthetic chemicals together are probably far less detrimental to wildlife species than habitat destruction resulting from encroachment by civilization and the burgeoning human populations.

TABLE 1-1:   Common Units

Metric Unit	Abbreviation	Equivalent to:
kilogram	kg	1,000 g, 1 million mg, 2.2 lbs
gram	g	1,000 mg, 1 million µg, ~0.035 oz
milligram	mg	1,000 µg, 1 thousandth of a g
microgram	µg	1,000 ng, 1 million pg
nanogram	ng	1 billionth of a g
picogram	pg	1 trillionth of a g
liter	L	~1 quart, ~33 oz
pound	lb	16 oz, 454.5 g, 0.45 kg
ounce	oz	28.4 g

TABLE 1-2:   Common Concentrations

Concentration	Abbreviation	Equivalent to:
milligram/kilogram	mg/kg	ppm, pg/g
microgram/kg	µg/kg	ppb, ng/g
nanogram/kg	ng/kg	ppt
milligram/liter	mg/L	ppm
parts per million	ppm	mg/kg, µg/g
parts per billion	ppb	µg/kg, ng/g
parts per trillion	ppt	ng/kg

Although man-made chemicals form a far smaller group than natural chemicals, they have become the symbols for the damage that the human species is inflicting on planet Earth and all its inhabitants. Why have synthetic chemicals been singled out for this distinction? One reason is that humans have been irresponsible, often unknowingly, in their use and disposal of synthetic chemicals, which are the relatively new products and tools of our civilization. As a result, problems of air, water, and general environmental pollution have been visited on societies throughout the world. Further, because synthetic chemicals are created by humans, there is a sense that they can be controlled by humans. The synthesis of new chemicals can be prevented, and the production of old chemicals can be halted.

A second reason relates to the general feeling that natural chemicals pose no threat. The theory is that humans and animals evolved with natural chemicals and are therefore adapted to them. This theory is not in accord with known adverse effects of natural chemicals in humans, such as the carcinogenicity of certain mold toxins or the acute toxicity of chemicals produced by a variety of microorganisms.

WHY THE "GOOD-BAD" DICHOTOMY?

What are the properties of synthetic chemicals that have fostered the dichotomy of man-made (bad) versus natural (good)? An exploration of this question is important to an understanding of the effects of chemicals on living organisms. The attributes of synthetic chemicals that set them apart from chemicals of natural origin were elegantly described years ago by the American biologist Barry Commoner:

> The clash between the economic success of synthetic petrochemicals and their increasing vulnerability to biological complaints is the inevitable result of the fact that they are synthetic-made by man, not nature. In every living cell there is a tightly integrated network of chemical processes which has evolved over three billion years of trial and error. In all of the countless organisms that have ever lived over this time, and in all of their even more numerous cells, there have been a huge number of opportunities for chemical errors—the production of substances that could disrupt the delicately balanced chemistry of the living cell. Like other evolutionary misfits, any organism that made these chemical mistakes perished, so that the genetic tendency to produce the offending substance was eliminated from the line of evolutionary descent. One can imagine that at some point in the course of evolution some unfortunate cell managed to synthesize, let us say DDT—and became a casualty in the evolutionary struggle to survive.
>
> Another requirement for evolutionary survival is that every substance synthesized by living things must be broken down by them as well be biodegradable. It is this rule which establishes the distinctive closed cycles of ecology. When petrochemical technology synthesizes a new complex substance that is alien to living things, they are likely to lack the enzymes needed to degrade the substance—which then accumulates as waste. This explains why our beaches have become blanketed in debris, since non-degradable synthetics have replaced hemp, cordage, wooden spoons and paper cups, which, because they were made of natural cellulose, soon decayed.
>
> The likelihood that a synthetic organic chemical will be biologically hazardous increases with its complexity; the more elaborate its structure, the more likely that some part of it will be incompatible with the normal chemistry of life.
>
> —THE PROMISE AND PERIL OF PETROCHEMICALS,
> *NEW YORK TIMES MAGAZINE*, SEPTEMBER 25, 1977, P. 38.

We can distill from this essay three attributes that make man-made chemicals biologically undesirable:

1. They are made by humans, not nature.
2. They are not biodegradable.
3. They tend have very complex structures.

None of these attributes bears any relationship to toxicity or the ability of a chemical to do harm. Let us examine each attribute individually to understand why this is true.

Man-made Chemicals Are Made by Humans

The first attribute—man-made chemicals are harmful because they are made by humans—is a commonly held opinion. It is an example of a form of reasoning that in logic is known as *circulus in probando*, or, literally, "a circle in the proof." Water is wet because it is water, man is human because he is man, and truth is good because it is truth are examples of circular reasoning. Such logic adds nothing to the argument that synthetic chemicals are biologically damaging. It returns us to the dichotomy of man-made (bad) versus natural (good) without adding anything to our knowledge of why the dichotomy exists or of its validity.

The hypothetical cell described above as perishing in some eon past because it committed the blunder of synthesizing DDT, a complex man-made organic compound with insecticidal properties, is worthy of a moment's reflection. This unfortunate cell is intended to serve as a dramatic, and perhaps whimsical, example of the undesirability of synthetic chemicals. However, the lamentable cell could never be more than a creature of fiction because its tale is founded in fancy rather than fact. The reality is that living cells do not just suddenly synthesize highly complex molecules. Complex natural molecules are the end product of many biochemical reactions occurring in well-coordinated sequence.

The building of a complex molecule by a living organism may be likened to the building of an automobile on an assembly line. At each step along the way, some small change or addition is made until finally, at the end of the line, an automobile emerges. So it is with complex biochemicals; each reaction in a biochemical chain makes some small change or addition to the molecule produced by the preceding reaction. This process is repeated numerous times, with each reaction producing the precursor for the next reaction in the chain, until a complex biochemical molecule is synthesized.

If all intermediary precursor biochemicals are compatible with cellular life, it is highly unlikely that the final step in the synthesis would produce a biochemical lethal to the cell. Thus, if the hypothetical cell described earlier did manage to synthesize DDT, it probably would be unaffected by the presence of DDT within itself. If the unfortunate cell were a plant cell, its destiny might be radically changed. Consider the tremendous survival advantage that a built-in insecticide would confer on a plant! The idea of a plant cell producing an insecticidal chemical is not as absurd as it may seem on the surface; pyrethrins and nicotine, produced by chrysanthemums and tobacco, respectively, are commercially available insecticides, and they function in the plant to control external pests.

The concept that chemicals are good or bad depending on their origin (nature or the laboratory) requires that chemicals possess an inherent moral quality of "goodness" or "badness," which, of course, is not true. Morality is a creation of the human mind and applies only to human conduct, not to inanimate things such as chemicals. The anthropomorphic view that nature, evolution, living cells, or inanimate objects are endowed with human sensibilities is out of context with reality. Such a view of cellular functions or natural processes is common among primitive cultures and medicine men. Scientists and authors may use the term *Mother Nature*, or credit natural processes with intelligence as a figure of speech for literary effect, but they do not invoke the gods to explain phenomena outside their areas of scientific expertise.

The distinction between natural and man-made chemicals is actually a man-made distinction. Living cells are not conscious units capable of deciding whether molecules that enter them are natural or synthetic. Our bodies cannot recognize the origin of a chemical—Mother Nature or the chemical laboratory. Our bodies can distinguish only between molecules they can use (for energy or to make more of themselves, more muscle, more bone, more blood, etc.) and molecules they cannot use.

Biochemicals may be natural or man-made, and foreign chemicals (xenobiotics) may be natural or man-made. The distinction between biochemicals and foreign chemicals exists for all living organisms and varies among classes of organisms. What may be a biochemical for one class of living things may be a foreign chemical for others. For example, strychnine is a natural chemical produced by *Nux vomica* plants. Thus it is a biochemical for *Nux vomica* plants but a foreign chemical (and a deadly one at that) for animal species, including humans. Although strychnine is a very toxic chemical for many species for which it is foreign, foreign chemicals are not of necessity harmful. The toxicity of chemicals does not correlate with their origin. Both natural and synthetic chemicals have wide ranges of toxicity, with large areas of overlap.

Man-made Chemicals May Not Be Biodegradable

The second attribute of man-made chemicals that allegedly makes them undesirable is their lack of biodegradability. *Biodegradation* refers to the process by which living organisms break down (metabolize) complex molecules to simpler molecules so that they can be eliminated by the body. In actual fact, there are relatively few man-made compounds that are not metabolized to some degree by some living organisms. All higher animals, including humans, have very complex sets of enzyme systems that process xenobiotic chemicals. These enzymes found primarily in the liver are referred to as the cytochrome P450 metabolizing enzymes. Among the synthetic compounds that are most resistant to biodegradation are the long-chain polymers that we know as plastics. As a class, the plastics are nontoxic. In fact, some are sufficiently inert biologically to be implanted surgically as substitutes for blood vessels, bone, and other living structures.

There are some substances produced in nature that are as resistant to biodegradation as synthetic polymers, such as the skeletons of diatoms, sea shells, bones, and hair. Thus, the bones of prehistoric humans and animals can be studied by archeologists today, thousands and thousands of years after their owners walked on Earth. And Napoleon's hair was still intact 140 years after his death, available to chemists for arsenic analysis to test the theory that he died of arsenic poisoning.

Chemicals, both natural and synthetic, that are resistant to biodegradation may be either toxic or nontoxic. If they are toxic, they enter the organism and may do damage because they are not converted to a less toxic form. If they are nontoxic, they enter and do no damage because they are not converted to a more toxic form via the metabolic pathways. In both cases, they are essentially unchanged by their passage through the organism. Chemicals, both natural and synthetic, that are biodegradable may also be either toxic or nontoxic. Some that are themselves nontoxic may be converted into toxic compounds during the process of metabolism, although this is relatively rare as the metabolism of xenobiotics *usually* leads to less toxic forms.

For these chemicals, biodegradability is a disadvantage to an organism; the process of biodegradation produces toxins. Other chemicals that are themselves toxic may be metabolically converted to less toxic or nontoxic compounds. For these chemicals, biodegradability is an advantage, and the process is called detoxification. Biodegradability and toxicity are independent properties of chemicals.

Nonbiodegradable junk may offend our esthetic senses when it clogs our beaches in debris, but plastic spoons and cups are no more esthetically offensive than wooden spoons and paper cups, which, if not picked up and discarded, have a long residence time as junk. If industry's successes in making biodegradable plastics widespread becomes a reality, the plastic counterparts may soon become more quickly degradable than paper or wood.

Most foreign objects, whether they are made of natural or synthetic materials, have the potential to harm any creature that ingests, inhales, or becomes entangled in them. The damage that has been done to some marine birds and mammals by plastic objects relates primarily to the form or shape of those objects. Children who play with thin plastic bags are also at risk if they put the bags over their heads and faces; they may suffocate because the plastic film does not permit air exchange. Plastic bags themselves are generally inert, but they can cause physical damage and even death.

Biodegradable plastics may not solve the ecological problem completely. A seal pup with its mouth held shut by a six-pack ring cannot wait a few months or even a few weeks for the ring to decompose. Changes in design of containers and packaging may be required as well as changes in the composition of the materials used. However, the problem of plastic litter is as much a societal problem as an industry one. We must accept the responsibility for proper disposal of debris that may harm some creature. It does not take much time to cut open six-pack rings or tie knots in plastic bags.

Further, the damage done by plastic debris to fish, fowl, and aquatic organisms, as well as to the esthetic beauty of our beaches, is small compared to that done by crude oil spilled from tankers, accidentally or deliberately, or released during blowouts from offshore wells. Crude oil, a raw material for synthetic chemicals, is not man made but produced by natural processes from once-living organisms. Some components of crude oil, all of which are natural, are notably resistant to biodegradation, as any resident of a coastal town plagued by an oil spill will attest. An example of a useful crude oil derivative that is resistant to biodegradation is asphalt, which is used to pave our streets.

A property of chemicals that results from a lack of biodegradability is persistence, the ability to remain in the environment unchanged by such factors as light, temperature, or microorganisms. Environmental concerns about persistence are related almost entirely to pesticides. Persistence is a desirable quality in pesticides, from the viewpoint of effectiveness and efficiency, since a pesticide that retains its ability to kill pests for prolonged periods need be applied less often than one that degrades rapidly. Thus, the total quantity of pesticide required to do a job is considerably less, which reduces the cost of crop protection and production.

The undesirable aspects of persistence in pesticides relate to their continued action after they are needed and to the fact that they remain in the environment for prolonged periods. The majority of persistent chlorinated hydrocarbon pesticides, such as DDT, have been banned from use in the United States because they were considered responsible for declines in the populations of certain wildlife species.

The rationale that substitution of nonpersistent pesticides for persistent ones will solve all of the environmental problems attributed to the latter is an example of the myopic thinking that permeates so many decisions relating to environmental protection. The rationale seems to be based on the notion that a nonpersistent pesticide does its job and then immediately, in a puff, dematerializes into nothingness. On the contrary, all nonpersistent pesticides merely degrade to other chemicals! The only difference is that most of these new chemicals do not have the same pesticidal action as their parent chemicals. These new chemicals may not kill pests, but what is their toxicity to other organisms? What is their fate in the environment? Do they persist? Do they accumulate?

A great deal of data indicate that some degradation products of nonpersistent pesticides have at least as much potential for nontarget damage as DDT. The identities of many of these degradation products are known because one of the requirements for registration of pesticides for commercial use is study of their environmental fate. However, there is absolutely no program for environmental monitoring of persistent products of nonpersistent pesticides. There is no demand from the groups that lobbied so hard to ban persistent pesticides to investigate the potential environmental damage from nonpersistent pesticides. Why? This is a philosophical question worthy of pursuit for anyone truly concerned about protection of the environment.

The demand for more applications of nonpersistent pesticides may result in an increased environmental burden of degradation products since the degradents themselves may be toxic. By forcing a ban on persistent pesticides, environmentalists may very well have created a much larger environmental problem than the problem they perceived as requiring the ban. Time will tell, if someone asks the right questions.

Time has already told us that the switch from persistent to nonpersistent pesticides has greatly increased the number of acute poisonings among farm workers. Cases of acute illnesses from chlorinated hydrocarbon insecticides were virtually nonexistent prior to the ban. The worst problems were cases of skin irritation. When DDT was banned, the use of organophosphate insecticides increased greatly. A large increase in poisoning of farm workers accompanied this increase; some poisonings were so severe as to be lethal. The efforts to protect wildlife had the ultimate effect of producing acute health problems among workers in the agricultural industry. Despite elaborate programs of worker protection—medical surveillance, protective clothing and cleanup, automatic measuring and mixing devices to avoid human contact, restrictions on reentry into treated fields, and so on—poisonings of farm workers by nonpersistent pesticides still occur. With the advent of a "natural" pesticide like BT (*Bacillus thuringiensis*, a bacteria that can be genetically engineered to enhance specific traits and is used to control gypsy moths), people hoped that poisonings would decrease. However, BT's usefulness is limited to certain classes of insects, and it is harmful to all butterflies and moths. Additionally, some chemically sensitive people believe that they respond just as severely to BT as to other pesticides.

Man-made Chemicals May Be Very Complex

The third attribute of man-made chemicals that presumably makes them undesirable is their complexity. Many synthetic organic chemicals do have very complex structures. The majority of synthetic chemicals are petrochemicals, which simply means that they are derived from petroleum. Petroleum is an organic substance; thus, petrochemicals are organic chemicals. Organic molecules, both natural and man made, may be very large and complex or small and relatively simple. Organic compounds are those having the element carbon in their structures. The combination of carbon atoms with atoms of other elements, primarily hydrogen and oxygen, gives rise to an extremely large group of compounds containing many subgroups with widely varying properties and uses. The orientation of the component atoms to each other is an important factor in determining the properties of the individual carbon compounds. All organic compounds, whether natural or synthetic, tend to be much more complex than inorganic compounds because of the nature of the carbon bond.

Some of the most complex chemicals that exist are those produced by living organisms, such as enzymes, hormones, and the deoxyribonucleic acid (DNA) molecules that carry genetic information. Some of these natural compounds have been synthesized in whole or in part in the laboratory. In fact, many natural organic compounds are so complicated that even the details of some of their structures remain hidden. Like nature, humans have also synthesized some very complex organic compounds, including the polymers, pharmaceuticals, and pesticides mentioned previously. With the advent of cell culture technology and the ability to insert the DNA of other species into bacterial cells, many complex hormones and enzymes can be synthesized in the laboratory. Although these are considered to be synthetic, they require the help of other organisms. The complex creations of both human and nature may or may not be toxic.

There are many relatively simple inorganic chemicals, such as cyanide and arsenic compounds, all of which may occur naturally or can be synthesized in the laboratory, that are much more toxic than complex synthetic compounds such as the drug aspirin or the pesticide malathion. Complexity and toxicity are independent properties of chemicals.

To understand what makes chemicals toxic, it is essential to recognize that the degree of toxicity of chemicals does not correlate in any way with whether they are natural or synthetic, biodegradable or nonbiodegradable, simple or complex. Arguments to the contrary, no matter how eloquently they are presented, have no basis in fact and serve only to confound the public.

CHAPTER *2*

WHAT HARM DO CHEMICALS CAUSE?

Toxicity is just one of the many ways by which chemicals can cause harm. Yet public concern about chemical exposures relates almost solely to the toxic properties of chemicals. This chapter helps broaden that view by describing briefly the other harmful properties of chemicals. Recognition of the fact that chemicals present many different kinds of dangers is essential to protect against human or environmental damage from chemical exposure.

HARMFUL PROPERTIES OF CHEMICALS

Explosiveness and Reactivity

Some chemicals are explosive when subjected to physical impact or high temperatures, when they come into contact with air or water, or when they combined with other chemicals. Explosions, depending on their severity, can cause injury or death as a result of physical or thermal damage to our bodies. Chemicals that are inherently explosive are not usually found in home settings, but certain chemicals that have a potential for explosion when placed under pressure are found in homes. Some of these chemicals are found in the ubiquitous aerosol canisters that dispense a wide variety of products—foods, hair sprays, pesticides, and so forth. The propellants used in spray cans are gases under ordinary conditions. When placed in pressurized containers, they become liquid and occupy a great deal less volume than they do in the gaseous state. When pressure is reduced by pressing the nozzle, they return to gas form, expand rapidly, and carry the product with them. Under normal conditions of use, aerosol products present little hazard of explosion, but if treated carelessly, they can become little bombs or missiles. Aerosol cans should never be subjected to sharp impact, punctured, placed near a high-heat source, or left for long periods in strong sunlight or very

The Dose Makes the Poison: A Plain Language Guide to Toxicology, Third Edition.
By Patricia Frank and M. Alice Ottoboni
© 2011 Patricia Frank and M. Alice Ottoboni. Published 2011 by John Wiley & Sons, Inc.

hot cars. Heat makes gases expand, greatly increasing the pressure inside the container, and can result in an explosion.

Other chemicals that present explosive dangers not generally recognized by the public are found in pressurized gas cylinders. For example, propane cylinders are commonly found in home settings. Some are designed for use with propane torches, and others are used with barbecue grills. Pressurized gas cylinders are also found in hospitals, laboratories, and industrial settings. All pressurized gas cylinders should be handled with extreme care to avoid high temperatures and sharp impact. They should be secured properly during storage and use. They are an especial danger in a fire.

Some chemicals are dangerous because of their chemical reactivity. For example, granular hypochlorite swimming pool chemicals are strong oxidizers. These chemicals can react spontaneously and cause a fire if they come in contact with oils or other combustible materials. In a fire, oxidizing chemicals make all other materials burn more violently. All swimming pool chemicals should be handled and stored with appropriate care.

Common household products provide examples of chemicals that produce a different type of danger as a result of their reactivity. Mixing chlorine bleach and household ammonia, a relatively frequent occurrence, can produce very harmful gases known as chloramines. Chloramines have a very low solubility in aqueous solutions; thus, they bubble out of solution rapidly when bleach and ammonia are mixed. Another dangerous combination is chlorine bleach and vinegar. The acid in vinegar causes rapid evolution of chlorine gas. The labels of household products warn of incompatible combinations and should be read and understood before the products are used.

Flammability and Combustibility

The distinction between flammable and combustible is based on ease of ignition. Flammables ignite at lower temperatures than combustibles. The Department of Transportation (DOT) uses makes this distinction for labeling of materials in transport. Some chemicals are flammable and will burn when they come into contact with a spark or flame at room temperature. Gasoline is an example of a highly flammable liquid. All of the petroleum distillates, such as kerosene, naphtha, and the various hydrocarbon solvents, are either flammable or combustible. They should be handled and stored in a manner that keeps them away from heat and flames. DOT requires that all hazardous materials being shipped carry a hazardous material (HazMat) sign designating the hazard. (See Figure 2-1(a)).

(a)   (b)

FIGURE 2-1   (a) Example HazMat sign; (b) HazMat sign for radioactive material.

The chlorofluoro hydrocarbon propellants used in most aerosol cans until recently are not flammable, but their replacements may be. Concern about the effects of chlorofluoro hydrocarbons on the ozone layer in the upper atmosphere has led to their replacement with other gases. The caution statement on the product label should tell if the propellant is flammable. If it is, protection should be taken against the flammability hazard as well as the explosion hazard. It is important that no spray stream contacts a lighted cigarette, pilot light, or other flame. Flammable chemicals cause injury or death as a result of tissue destruction from thermal burns.

Radioactivity

Radioactivity is the spontaneous emission of energy as particles or rays. Naturally occurring and man-made elements may be radioactive. Because radioactive elements retain their chemical properties, they can react to form radioactive chemical compounds that have the same chemical properties as their nonradioactive counterparts. Many elements exist in both radioactive and nonradioactive forms. An example is potassium. Although it is not commonly considered a radioactive element, about 0.01 percent of all the potassium on Earth is radioactive; the remainder exists as nonradioactive isotopes.

Elements with atomic numbers higher than 82 occur predominately or entirely in their radioactive forms. Some commonly mentioned examples of this group are uranium, radium, and radon gas. Radioactive chemicals cause damage primarily by emission of subatomic particles or rays known as alpha particles, beta particles, and gamma rays. These emissions can alter bio-chemicals that they strike within cells. The damage done depends on the molecules that are altered. Interactions with DNA molecules can produce mutations (see Chapter 7). Radioactive chemicals act biochemically in the same as any other chemicals. They may enter the body, where they may be metabolized, excreted, or stored depending on their chemical properties. They are particularly dangerous inside the body, because radiation emitted by these chemicals can directly damage cells and tissue. A good example is radium. Body metabolic processes will store radium in bone tissue. Stored radium continuously bombards adjacent bone tissue. Figure 2-1(b) shows the HazMat sign for a radioactivity hazard.

Studies of luminous watch dial painters who through job exposure ingested small amounts of radium show that radiation from the stored radium resulted in leukemia, a form of cancer. The way the radium was ingested is an interesting worker safety story. Dial painters in the 1920s (mostly women) needed a sharp point on the very fine brushes they used to apply the radium. They placed the brushes in their mouths or licked them to make the point sharp. The workers were never warned about this hazard since this was before we knew about the dangers of radioactivity. Because cancer, as we will learn in Chapter 7, has a long induction time, the cause-and-effect relationship between ingesting radium and leukemia as well as other forms of cancer was not immediately obvious.

The sources of exposure to radioactive chemicals can be placed in four groups:

1. Naturally occurring radioactive elements, such as uranium, radium, and radon

2. Medical application for diagnosis or treatment, such as radioactive iodine cocktails for study of thyroid function or thallium for heart imaging

3. Products of nuclear reactions, such those from nuclear power plants or nuclear explosions

4. Miscellaneous industrial or research applications, such as use of radioactive compounds as biochemical markers in the study of the metabolic fate of chemicals

Radioactivity from external sources, such as X-ray machines, computed tomography (CT) scans, positron emission tomography (PET) scans, or other radiation sources, can also be harmful. The important point is that these external sources present different problems and risks from the internal sources ingested or injected in the form of radioactive chemicals. Protection against external sources may be accomplished with shielding and safe distance. Protection against internal sources requires protection against the entry into the body of radioactive materials from contaminated air, water, food, and environment. The biochemistry and physics of radioactive materials are complex subjects. For more information, reference materials from a local library or online such as the site of the physics department from Idaho State University (www.physics.isu.edu/radinf/natural.htm) give a more in-depth discussion of radioactivity.

There has been a great deal of publicity about the presence of radon in homes and other buildings. Radon is a radioactive gas that is given off by soil, rocks, and building materials made from them, such as concrete and stone. The source of radon is radioactive decay of uranium, which is distributed throughout Earth's crust. The amount of radon emitted from the Earth varies with the kind of rock formation present and with the geographical location. Granite and shale contain more uranium than other kinds of rocks and thus emit more radon. The radon accumulates in buildings because structures have poor ventilation. Radon does not accumulate and present a hazard in outdoor air.

The known adverse effect from exposure to radon and the radioactive decay products formed from it is lung cancer. People who want more information about detecting and controlling radon exposures in homes or workplaces should contact their regional office of the Environmental Protection Agency or go online at www.epa.gov/radon for pertinent information.

Corrosiveness

Some chemicals are corrosive and do damage by destroying tissues that they contact. They can destroy skin, but are especially destructive to eyes, mucous membranes of the mouth and throat, and linings of the lungs, esophagus, and stomach. Examples of common corrosives are sodium hydroxide (caustic soda, lye), potassium hydroxide (caustic potash), hydrochloric acid (muriatic acid), and sulfuric acid (oil of vitriol). Corrosive chemicals, like flammable chemicals, cause injury or death as a result of tissue damage. The difference between the two is that corrosives cause chemical burns whereas flammable chemicals cause thermal burns.

Irritation

Some chemicals are irritants. They do not destroy tissues as do the corrosives, but they produce varying degrees and combinations of redness, swelling, blistering, burning, or itching sensation. Dilute solutions of corrosives, many solvents and polishes, liquids with a low or high pH (acidity or alkalinity), turpentine, pine oil, and spices such as nutmeg, cinnamon, pepper, mustard, and clove are examples of irritants. As a general rule, skin contact with irritant chemicals does not cause death or permanent injury as a result of irritant properties; however, their effects can be very discomforting, both physically and cosmetically. Inhalation of irritant gases such as nitrogen oxides, or irritant fumes such as cadmium oxide, can cause pulmonary edema, which can be fatal.

Additionally, the eye may be more sensitive to the irritating effects of chemicals and exposure may result in blindness. This is why the ingredients in cosmetics need to be tested for both skin and eye irritation; cosmetics frequently come in contact with the eye during application or from transfer from the hand to the eye. Hair spray, spray-on deodorant, and feminine hygiene sprays are the common spray products that can get into the eye and frequently cause irritation. Cosmetics that claim they have never been tested in animals contain chemicals that were previously tested and found to be nonirritants. The use of a chemical in cosmetics that has not *ever* been tested for irritation would be dangerous if not actually illegal in the United States.

Sensitization and Photosensitization

Some chemicals can harm us by causing a sensitization reaction. Sensitization is an allergic response. The subject of allergy, a subdiscipline of the science of immunology, is a complex topic. We include a brief and oversimplified description of the sensitization reaction, since this is one of the more common mechanisms by which chemicals cause harm.

Chemicals that cause sensitization are called allergens (or antigens). A single exposure or, in some cases, several repeated exposures to an allergen by inhalation, skin contact, or ingestion may cause the body of the person exposed to manufacture antibodies that will react with the allergen. Once antibodies are formed, the person is sensitized, after which any future exposure to the allergen, even in amounts very much smaller than the original sensitizing dose or doses, results in an allergic reaction. Allergies may manifest themselves by such symptoms as dermatitis, hives, red or puffy eyes, sneezing, runny nose, headache, and asthma, singly or in combination. In severe cases, allergies cause anaphylaxis, a severe hypersensitivity reaction.

Sensitization is an immune system response. The complexity of immune system reactions is demonstrated by the fact that another response, immunization, involves a sequence of events similar to that of the sensitization response. However, unlike sensitization, which causes illness, immunization *protects* against illness. Immunization is the process whereby protection against harmful microorganisms, such as smallpox, measles, influenza, or rabies, is obtained. In immunization, harmful organisms (antigens) that have been rendered incapable of causing symptoms of disease but not incapable of being recognized by the body's immune system are injected into the body. The body in turn makes antibodies against the antigen. Once antibodies are formed, the person is immunized. Any future exposure to the harmful organisms results in a reaction between the organisms and the antibodies. This reaction kills or inactivates the organisms, thus preventing disease. Immunization may be considered an appropriate response of the immune system; sensitization, however, is an inappropriate response.

It is probable that any chemical, natural or synthetic, is capable of causing an allergic reaction in some individual somewhere in the world, but some chemicals cause sensitization in a significant number of the people with whom they come in contact. Examples of such substances are pollens of all varieties, poison ivy, poison oak, orris root, epoxy resin components, and formaldehyde. Orris root was used many years ago as a base for face powder until its allergenic properties were recognized. Formaldehyde has long been recognized as a sensitizer in occupational settings, but it was not known that it was a problem for the general public until it was identified as a component of indoor air pollution. (See Chapter 9.) People who are sensitive to formaldehyde should know that permanent press clothing may contain trace amounts of the chemical, which remain as a residue from the treatment process. A few washings should remove this residue.

For most people, sensitization is an annoyance and a discomfort, but for some people, it is very debilitating and may even result in radical changes in work, lifestyle, or place of residence in order to avoid trace exposures to the offending allergens. Fortunately, deaths from an exaggerated allergenic response (anaphylactic shock also called anaphylaxis) are rare.

There has been an enormous increase in the number of children allergic to ground nut (peanut) products. The reasons for this increase are not currently known, but reactions to peanuts in allergic people can be severe. The Mayo Clinic Web site (www.mayoclinic.com/health/peanut-allergy/DS00710) has information on this phenomenon. Recent experiments have shown that it is possible to desensitize some people so that they do not have a full allergic response. "Death by peanut" is a rare occurrence, although each event is surrounded by so much publicity that it appears more common.

The most common allergy to any medicine is probably to penicillin. Numbness or tingling of the extremities and face, breathing problems, and anaphylaxis are the common symptoms of penicillin allergy. Patients who have ever experienced these symptoms should not be given penicillin again and should be sure to inform their medical caregivers of this allergy since penicillin is commonly used. Again, desensitization to penicillin can be accomplished in some individuals following a careful program administered under a physician's supervision.

Sensitization must be distinguished from irritation, which it can mimic, particularly when the skin is involved. Irritation and sensitization can both manifest themselves as dermatitis, an inflammation of the skin. Irritants can also produce the same eye symptoms as allergens if rubbed in the eye or chest symptoms if the irritant is inhaled. Irritation is a purely local or topical phenomenon, whereas sensitization is a systemic condition involving the whole body. Only a physician can determine if symptoms are allergic in nature.

Photosensitization is similar to sensitization, except that light is required to trigger the adverse reaction. Photosensitization may occur as a result of oral, dermal, or inhalation exposure. The photosensitization reaction has several manifestations. There may be hivelike swellings, acnelike eruptions, severe sunburn, or a combination of the three in areas of the skin exposed to light. With repeated episodes of photosensitization, the affected skin may develop a permanent darkening and be more subject to skin cancer. Examples of chemicals that can cause photosensitization are coal tar compounds, such as creosote and dyes derived from them; some forage plants, such as St. John's wort and buckwheat; and a number of drugs, such as phenothiazine, aureomycin, and sulfonamides. Severe phototoxicity may produce a syndrome called Stevens-Johnson syndrome, an immune reaction involving the skin and mucous membranes of the eyes, mouth, lungs, and other organs that can lead to disease and death. Some medicines have a warning label about this condition, and it is important to stay out of the sun if you are taking these medicines.

Toxicity

Finally, many chemicals can cause harm by virtue of their toxicity. Toxicology is the science that investigates the adverse effects of chemicals. The mechanisms by which a chemical exerts a toxic action are many and varied. Some chemicals are themselves toxic. Others must be converted metabolically to other forms before they become toxic. Some chemicals may damage any cells or tissues with which they come in contact. Others damage only one specific kind of cell or tissue. Some chemicals act directly on cells or tissues. Others act indirectly in some manner, such as by interfering with some biochemical reaction, altering some physiological mechanism, or destroying an essential nutrient, which in turn is the damaging event. Regardless of the mechanism, the principles that govern the harmful property of chemicals known as toxicity apply equally to all chemicals, both natural and synthetic.

The toxicity of a chemical refers to its ability to damage an organ system, such as the liver or kidneys, or to disrupt a biochemical process, such as the blood-forming mechanism, or to disturb an enzyme system at some site in the body removed from the site of contact. Toxicity is in contrast to corrosiveness, which does damage at the site of contact. The systemic damage that chemicals do is not capricious or random. A chemical does not affect one set of functions in one person and a different set in another. If it did, its effects would be unpredictable, and there could be no such thing as a science of toxicology. A chemical can have an effect on several different functions within an individual, and individuals can vary among with regard to the sensitivity of different functions. Thus, for example, it may appear that a chemical affects the kidneys in one person and the liver in another, but a detailed examination such as at autopsy will reveal that both liver and kidney were involved, but different genetic makeup made one organ system more susceptible.

Every chemical, synthesized or naturally occurring, has some set of exposure conditions in which it is toxic. Conversely, every chemical has some set of exposure conditions in which it is not toxic. Chapters 3 and 4 explore this concept in depth.

Multiple Harmful Properties

Some chemicals may possess only one of the harmful properties just discussed, whereas others can produce damage by several means. For example, hydrochloric acid and sodium hydroxide are corrosive in concentrated form and irritants in dilute solution. Neither is classed as toxic—that is, neither cause damage at a location removed from the site of contact—but both can cause death by destroying tissues they touch. In such cases, death is the result of the body's inability to survive the harm caused by tissue destruction.

Gasoline, while extremely flammable, has a relatively low acute toxicity by itself. It presents a unique danger if, when ingested, some of the liquid is aspirated. Gasoline and all other petroleum distillates when taken into the lungs in liquid form will produce a chemical pneumonitis, that is, an inflammation of the lung tissue. The fatality rate from petroleum distillate pneumonitis is quite high. For this reason, household products that contain 10 percent or more of petroleum distillates must carry a warning not to induce vomiting if ingested. The act of vomiting greatly increases the chances that some of the material will be aspirated. Petroleum distillate pneumonia is seen most frequently in children who drink furniture polish. However, adults also can become very seriously ill or die as a result of accidentally aspirating gasoline, such as can occur while siphoning it by mouth from an automobile gas tank.

The only other chemicals that are required by law to carry a "Do not induce vomiting" warning on the label are the corrosives. With corrosives, the danger is not chemical pneumonitis but rather the chance that the act of vomiting will subject a corroded esophagus to sufficient stress to cause it to rupture.

DEFINITION OF *POISON*

A chemical that causes illness or death is often referred to as a poison. This concept is erroneous and has resulted in a great deal of confusion in the public mind about the nature of the toxic action of chemicals. *Poisons* are chemicals that produce illness or death when taken in small quantities and that act systemically. As we mentioned before, any chemical can be toxic when taken in a large enough dose, but this is not the definition of a poison. For example, water when ingested at greater than four to six liters at one time may be toxic or even lethal, but water is not a poison.

DEFINITION OF *HAZARD*

Another distinction that should be made is the difference between *toxicity* and *hazard*. The latter term has come into common use as a synonym for the former. Actually, hazard is a much more complex concept than toxicity because it includes conditions of use. The hazard presented by a chemical has two components:

1. The inherent ability of a chemical to do harm by virtue of its explosiveness, flammability, corrosiveness, toxicity, and so forth
2. The ease with which contact can be established between the chemical and the object of concern

These two components together describe the chance that a chemical will do harm. For example, an extremely toxic chemical such as strychnine, when sealed in an unopenable vial, can be handled freely by people with no chance that a poisoning will occur. Its toxicity has not changed, but it presents no hazard because no contact can be established between the chemical and people.

FIGURE 2-2 "Mr. Yuk" poison sign.

Boric acid, which is a very effective roach killer, does not present a hazard when applied in powder form in spaces behind cabinets and under drawers where it is inaccessible. It can be very hazardous when combined with sugar and formed into tablets that are scattered around baseboards. A young child crawling around the kitchen floor can find and eat these tablets without being observed. Several hours later, when symptoms appear, the cause will not be known, and the urgency of getting medical attention will not be recognized. The fatality rate for accidental boric acid intoxication in children is about 50 percent.

Figure 2-2 shows the "Mr. Yuk" signage that is placed on poisons or other hazardous chemicals in order to warn children to avoid these products. Parents should teach their children to recognize Mr. Yuk and stay away from him.

CHAPTER 3

WHAT IS TOXICOLOGY?

Toxicology is the study of the adverse systemic effects of chemicals. All substances that caused illness or death were originally referred to as poisons. Today, the more general term *toxicant* is used, with *poison* being reserved for the special class of toxicants that require only small amounts to cause death. A toxicant now is defined as a chemical that does systemic damage, but the word came originally from the Greek words *toxon* and *toxikos*. Although the word *toxikos* or *toxicos* means "poison" in Greek, these words originally referred to bows and arrows used for hunting game. Early cultures found that their weapons were more effective if the arrows were dipped in toxic juices obtained from plants. Arrows coated with toxic juices would kill or immobilize the victim. If an arrow did not kill the animal or bird (or enemy), it could escape. The possibility that traces of arrow toxins might be present in the flesh of meat apparently did not concern primitive hunters. Meat obtained in such a manner today would probably not be accepted by the public, even if it did meet current regulatory requirements. The "-ology" portion of toxicology comes from the Greek for "study" or "knowledge." Thus, *biology* is the study of living things, *geology*, the study of Earth, and so on.

EMPIRICAL TOXICOLOGY

Empirical toxicology, as distinguished from the science of toxicology, predates recorded history. Prehistoric people knew which poisonous plants and animals to avoid from watching their fellow beings and by being educated by group elders. These experiences also told them which plant juices to use as toxins. The fact that certain substances could cause acute illness or death was known to even the earliest human populations because it was so obvious—the cause-effect relationship was so direct.

Before the beginnings of recorded history, a large number of plant and animal toxins were recognized. The early Greek physicians cataloged many

The Dose Makes the Poison: A Plain Language Guide to Toxicology, Third Edition.
By Patricia Frank and M. Alice Ottoboni

toxins—animal, plant, and mineral. As civilization progressed, so did the art of poisoning. Empirical toxicology became applied toxicology. Notable historic examples of the application of toxicology are the execution of Socrates by means of a hemlock brew, the suicide of Cleopatra by means of a bite of an asp, and the many political assassinations of ancient and medieval times by means of a variety of plant and mineral toxicants.

Empirical toxicology is concerned almost exclusively with the acute toxicity—the immediate toxicity—of chemicals. The fact that substances could cause chronic intoxication, either illness or death from long-term, low-level exposure to chemicals, was not recognized until relatively recently. The first hints that chronic exposure to chemicals could cause illness came from occupational exposures to mercury and lead.

Early Greek and Roman physicians noted the poorer health of people who worked at certain trades, particularly those involving manual labor, such as mining, metallurgy, or the making of pottery. Hippocrates described severe colic in men who extracted metals, which he attributed to lead poisoning. Pliny wrote of mercury poisoning among slaves who worked in the quicksilver mines of Almaden, Spain. But in those days manual laborers and slaves were considered to be inferior beings who lived unhealthy lifestyles, so their poor health was not considered remarkable. They suffered a high background incidence of disease, physical infirmity, and deformity caused by accidents, abuse, and the stresses of poverty. The diseases caused by the chemicals to which they were exposed blended into the background and were hidden. It was not for another 1,500 to 2,000 years that the connection between disease and chronic exposure to chemicals would be given serious consideration.

PARACELSUS AND RAMAZZINI

The first monograph on occupational diseases of miners and smelters was published in 1567, 26 years after the death of its author, the Swiss physician Phillippus Aureolus, who took the pseudonym of Theophrastus von Hohenheim and later was known as Paracelsus. In his monograph, Paracelsus distinguished between acute and chronic toxic effects of metals and described in detail the symptoms of chronic mercurialism.

Paracelsus was a man far ahead of his time, a bridge between alchemy and science. He was an iconoclast with contempt for the medical doctrines and methods of the day. Thus, he earned the disdain of the medical establishment, which led him to move his medical practice around Europe a number of times. However, he was an itinerant physician probably as much by choice as by necessity. Although he was disliked by his peers, Paracelsus was extremely popular with his students and patients, among whom were counted members of many of the ruling families of medieval Europe.

Paracelsus lectured and wrote not in Latin, which was the custom of the time, but in his native German. He preferred the company of laborers, tradespeople, Gypsies, and others unacceptable to genteel folk. He died at the age of 48 as a result of wounds suffered in a tavern brawl in Salzburg. In his short lifetime he set the stage for a revolution in medical practice by teaching his student physicians to use chemical medications rather than the more popular magic potions.

Paracelsus set forth one of the basic tenets of modern toxicology when he wrote what is translated as:

> What is it that is not poison? All things are poison and nothing is without poison. It is the dose only that makes a thing not a poison.

Testimony to this truth is found in Paracelsus's own writings. He not only was the first to publish a description of the symptoms of chronic mercury poisoning but also was the first to design a rational chemotherapy using mercury for the treatment of syphilis. Paracelsus's treatment was used until the discovery of the arsenic compound arsphenamine (Salvarsan) by Paul Ehrlich, a German bacteriologist, more than 350 years later. Thus, Paracelsus is considered one of the fathers of toxicology, but he was also one of the fathers of modern medicine.

Since the awareness that chemicals could be chronically toxic was born in the study of the diseases of occupations, no discussion of the origins of that awareness would be complete without mention of Bernadino Ramazzini, who is considered to be the father of occupational medicine.

Ramazzini was born in Italy in 1633 and lived to the age of 81, active in his profession until the end. Ramazzini's specialty was epidemiology, the study of the incidence, prevalence, and movement of diseases in a population (i.e., diseases that attack many people in a region at the same time). His book, *De Morbis Artificum Diatriba* (Diseases of Workers), listed the illnesses and diseases occurring in many occupations. He was also a socially oriented physician who noted the wretched conditions of workers in the trades. He felt that medicine should strive to improve the quality of life of the common people.

Ramazzini was once told by a sewer worker that a man who had never done such work could not possibly understand what it was like. Ramazzini took this admonition to heart. In all of his studies of the occupations, he followed his subjects into the mines, factories, and cesspools to see and experience the conditions under which they labored. He learned that there is no substitute for firsthand knowledge and that occupational exposures could be involved in the causation of disease. Ramazzini is remembered today by the medical profession for his wise counsel that, in addition to the usual questions, doctors should inquire into the occupations of their patients. He also used the bark of the cinchona tree to treat malaria. He was ridiculed by his peers, but, as it turns out, cinchona tree bark is rich in quinine, which became the first "modern" treatment for malaria.

A BRIEF HISTORY OF TOXICOLOGY

Pharmacology (from the Greek, *pharmakon*, "drug") is one of the oldest of the biological sciences. It is defined as the study of the action of drugs, their nature, preparation, administration, and effects. Substitution of the word *chemicals* for *drugs* in the definition would give a more appropriate description of the early science. The whole universe of known chemicals came under the scrutiny of early pharmacologists in their search for substances that had healing powers. All substances were considered to be potential candidates for medicinal use. Thus, the pharmacology literature is replete with information on beneficial and adverse effects of a tremendous number of naturally occurring chemicals, both organic and inorganic, that are not classed as drugs in the modern usage of the term.

The development of synthetic organic chemistry freed pharmacologists from some of their dependence on nature's bounty and enabled them to tailor-make drugs to fill specific therapeutic needs. Modern pharmacology, which developed along with modern medical science during the nineteenth and early twentieth centuries, has added many synthetic chemicals to its list of drugs and deleted most of the old-fashioned remedies from its inventory of medicinal preparations.

The science of toxicology, in contrast to empirical toxicology, is a young science compared to many other biological sciences. It was born of pharmacology and, until the mid-twentieth century, was considered a subdivision of the science of pharmacology. Pharmacology embraced toxicology because it is neither possible nor practical to study only therapeutic effects to the exclusion of toxic effects. In fact, the relationship between toxic effect and therapeutic effect forms the basis for an extremely important medical concept: the therapeutic index. The therapeutic index of a drug is the number obtained by dividing the dose that is toxic by the dose that gives the desired curative effect. The larger the therapeutic index (the greater the difference between the dose that harms and the dose that helps), the safer the drug.

A pharmacologist is a person who studies the action of drugs on people and animals at therapeutic levels. A pharmacist is a person who compounds and dispenses drugs. A toxicologist is one who studies the effects of chemicals across the range of doses from therapeutic to dangerously high.

Early interest in the toxic effects of chemicals was directed toward symptoms of acute poisoning, methods of treatment of poisoning victims, and later the legal aspects of criminal cases. The early Egyptian and Greek physicians recorded the effects of many poisonous plants and substances derived from them. Application of this knowledge became a way of life among the ruling Italian families during the fifteenth century. During the Renaissance, the act of poisoning was developed into a fine art that was practiced on relatives, friends, and political rivals. Borgia, the name of the notorious family that helped perfect the art, became infamous as a synonym for those who poison. The Borgias specialized in the use of arsenic but had a large armamentarium of poisons at their command.

Impetus for the study of toxicology as a science separate from pharmacology came primarily from occupational medicine. The first systematic classification of all of the chemical and biological information available at the time was published in the early nineteenth century by M. J. B. Orfila, a Spanish physician who taught at the University of Paris. History credits Orfila with being the first to consider toxicology as a discipline separate from pharmacology.

The rapid growth of the synthetic organic chemical industry during the mid- to late twentieth century brought with it a greatly heightened awareness and concern about the toxic action of chemicals. The science of toxicology is now fully recognized as a discipline separate from pharmacology. Toxicologists have their own vetting groups. Those passing the exam of the American Board of Toxicology are referred to as Diplomates (DABT), and those qualified by the Academy of Toxicological Sciences are referred to as fellows of the ATS. These qualifications are similar to those used by physicians to demonstrate proficiency in their specialties such as oncology and neurology.

In the early twentieth century, science and medicine were taking giant leaps forward with the discovery of human blood types by Karl Landsteiner, immunology experiments by Paul Ehrlich that led to the development of cancer chemotherapy, the isolation of insulin from the pancreas by Frederick Banting and Charles Best, and the discovery of penicillin by Alexander Fleming in 1928. By the mid-twentieth century, Willem Koliff developed the first kidney dialysis machine, James D. Watson and Francis Crick identified the structure of DNA, Jonas Salk developed the polio vaccine, and ultrasound was beginning to be used for medical diagnosis. Science was moving at a fast pace, so fast that products and drugs may not have been tested fully nor were people looking at the downstream effects of by-products of our stepped-up industrialization. It was into this mix of new but uncertain times that the science of toxicology was born.

The regulation of toxic chemicals in the United States predates by many decades current demands for control of synthetic chemicals that enter the environment. In the late nineteenth and early twentieth centuries, publicity about gross contamination and adulteration of foods that resulted from industry malpractices created a public demand for reform of the food industry. In addition, the public became concerned about chemicals that were added, either deliberately or inadvertently, to canned and packaged foods. Upton Sinclair's book *The Jungle*, published in 1906, graphically described the conditions in the meatpacking industry, leading to a public outcry primarily about food quality even though its author had written the book in order to arouse sympathy for the workers. This book is still a great read (and still scary).

During the same period, Harvey W. Wiley, the chief of the Bureau of Chemistry of the U.S. Department of Agriculture (USDA), was a vocal advocate of wholesome, unadulterated food. During the last quarter of the nineteenth century, Dr. Wiley and his staff studied the composition of foods and their adulterants. He was quoted as saying in 1902:

> There has been too much argument about the effects of chemical preservatives on health. I propose to find out by scientific experimentation what is the truth about a question of such vital concern to consumers of the nation. Some day we will have a law.
> —F. B. Linton, "Federal food and drug leaders,"
> *Food Drug Cosmetic Law Quarterly* 4 (1949): 451–470.

Dr. Wiley's prophecy came to pass when, after more than 20 years of consideration of similar measures by Congress, the Pure Food and Drug Act of 1907 was enacted. Administration of the act was placed in the Bureau of Chemistry, and Dr. Wiley was appointed the first Food and Drug Commissioner by President Theodore Roosevelt. His task was not easy. The Food and Drug Commissioner had authority to investigate but little power to enforce. Dr. Wiley and his assistants, known as the poison squad, made many political and industry enemies. However, it was too late to still public interest and awareness of the problem of contaminated and adulterated foods and drugs. Later, Dr. Wiley became the head of the *Good Housekeeping* magazine labs, which were responsible for testing consumer products. He died on the anniversary of the signing of the Pure Food and Drug Act in 1934.

In 1910, the Federal Insecticide Act was passed. It was the U.S. government's first attempt to control the sale and use of insecticides. The legislation did not stimulate much enforcement activity because so few insecticides were available for use by agriculture at the time. One of the most commonly used insecticides was the very toxic lead arsenate. Cases of accidental poisoning and acute illnesses among those who applied lead arsenate, combined with concerns about chronic health effects of its residues, kept interest in toxic chemical regulation alive, if relatively inactive, in the minds of legislators and regulators.

Two important events took place in 1927. First, increasing public demands for pure foods and drugs resulted in the creation of the Food and Drug Administration (FDA) to administer the Pure Food and Drug Act. All responsibility for the act was transferred from USDA to FDA. For the first time since 1907, an organization existed for the sole and specific purpose of enforcing pure food and drug laws.

The second important regulatory event of 1927 was passage of the Federal Caustic Poison Act. The many tragic deaths and disfiguring, disabling injuries among children who swallowed lye (sodium hydroxide) provided the stimulus for passage of this Act. Soap making was a common household chore for women of the time, particularly in rural communities. Lye, a necessary ingredient in making soap, was commonly found in cabinets accessible to small children. The act, which was primarily a labeling law, was an attempt to protect children from corrosive materials by requiring warning labels on such products. The labels informed adults of the dangers of ingesting these materials. Theoretically, the adults, in turn, would store the labeled products out of the reach of children.

Over the next years, few changes were made in the food, drug, and insecticide laws. A major regulatory action usually occurs only after a disaster or public demand requires it. Such was the case with the enactment of the Food, Drug, and, Cosmetics Act (FDCA) of 1938: Congress had been at work for years on measures to strengthen the 1907 act. The elixir of sulfanilamide tragedy of 1937 (see Chapter 6) provided the extra bit of incentive required for passage of a stronger and more protective law than the 1907 act.

Despite many decades of public concern about the safety and purity of foods and drugs, the modern era of toxic chemical regulation did not begin until after World War II. Legislation regulating toxic chemicals has for the most part followed public concerns. As mentioned, laws affecting foods, food additives, and cosmetics were passed in the first part of the twentieth century. After World War II, concurrent with the rapid development of the synthetic pesticides, laws were passed to control both their use and their residue levels in food. The public, legislators, and regulators recognized that advances in pesticide technology, which led to a tremendous increase in numbers of pesticides available and their volume of use, required increased public and consumer protection. The Federal Insecticide, Fungicide, and Rodenticide Act (FIFRA), enforced by USDA, was passed in 1947, and the Miller Amendment to the FDCA, enforced by FDA, was passed in 1954. In 1958, the much-publicized Delaney Clause—an amendment that prohibits the use of any food additive that is found to induce cancer when ingested by humans or animals—was added to FDCA. The provisions of FIFRA and FDCA supplemented each other. They provided uniformity in requirements for toxicity testing and tolerance-setting procedures for the chemicals under their respective purviews.

During the 1950s and 1960s, there was a steady increase in legislation regulating a wide variety of chemicals that enter homes, workplaces, and the general environment. During this period, public interest was expanded to include air and water pollution. A series of laws were passed to control automobile exhaust emissions and industrial discharges to both air and water. At about the same time, the need for environmentally sound methods for disposing of household, commercial, and industrial refuse became increasingly important. Problems associated with dumps—environmental damage, air pollution, ground and surface water pollution, and public nuisance—have resulted in a series of laws to recycle, treat, store, and dispose of both chemical and household waste.

Concern for the health and safety of children also held the attention of the public. In 1960, the much broader Hazardous Substances Labeling Act (HSLA) superseded the Caustic Poison Act of 1927. HSLA, like the Caustic Poison Act, was a child-protection measure. It applied to all products sold for home use and products that could conceivably be brought into homes because of small container size that are not covered by other laws, such as foods, drugs, pesticides, and cosmetics. HSLA, like the Caustic Poison Act, was only a labeling law. It defined many more categories of hazards associated with household products and specified labeling for each. The act was based on the assumption that parents will read labels and take appropriate steps to protect their children. HSLA was strengthened in 1970 by the Poison Prevention Act, which required that child-proof containers be used for products considered to be too hazardous for labeling alone to be effective in protecting children.

The Environmental Protection Agency (EPA) was established in December 1970. Rachel Carson's book *Silent Spring*, published in 1962 and reissued in 2002, stimulated public awareness of environmental chemicals, particularly pesticides. During the years that followed, the public became increasingly fearful that pesticides and other chemicals were harming them and the environment. The decision to bring regulation of the many kinds of environmental chemicals together in one agency, stimulated by public concerns about toxic chemicals, resulted in the creation of EPA.

Essentially all control over pesticide use was taken from USDA and FDA and placed in EPA. In addition, EPA was given jurisdiction over a wide variety of other chemicals that were potential pollutants in order to protect wildlife and the environment. In the 1960s and 1970s, interest also had begun to focus on chemicals used in industry and their impact on workers and on the neighborhoods surrounding individual plants. Expanding knowledge and awareness of chemical hazards provided a stimulus to industry for self-regulation. As a result, this period saw increased development of guidelines for procedures and practices to protect workers in chemical and related industries.

The first legislative action in response to the interest in worker health was the Occupational Safety and Health Act of 1970. Responsibility for enforcement was placed in a new agency, the Occupational Safety and Health Administration (OSHA), located in the Department of Labor (DOL). The act is a worker-protection law designed to prevent traumatic injuries as well as occupational diseases. It addresses the whole issue of job-induced injuries, including health hazards and their control. Its passage finally gave official recognition to the long-known fact that the people at great risk from exposure to toxic chemicals are the people who work with them.

One of the most important new ideas to come out of this law was the Hazard Communication Standard promulgated by OSHA. The Hazard Communication Standard, also called the worker right-to-know law, requires manufacturers and importers of chemicals to determine the hazards of each product they sell. They must also transmit the hazard information and associated protective measures to their employees and customers through labels and Material Safety Data Sheets (MSDSs). Employers, in turn, are required to make MSDSs available at job sites and to inform workers about the nature of the hazards involved and how to protect themselves. Workers have the right to see and study these data sheets.

The right of people to know about the chemicals in and around their own communities was brought into focus in December 1984 by a chemical plant disaster in Bhopal, India. Operating problems at this chemical plant resulted in the release of thousands of pounds of methyl isocyanate, a toxic and very corrosive gas, without warning into the nearby city of Bhopal. Approximately 2,000 people were killed and 30,000 were injured. After studying this accident and other plant disasters, the U.S. Congress passed the Emergency Planning and Community Right-to-Know Act of 1986. Among other important provisions, this law requires plant owners to notify specified government agencies of chemical inventories and chemical releases to air, water, and waste disposal facilities. It also requires plant owners to provide a chemical inventory list and copies of MSDSs to local fire departments. Under the law, all of these data are available to the public.

In 1927 the Bureau of Chemistry was renamed the Food, Drug and Insecticide Administration and came into being as the FDA in 1931. In 1938, the modern FDA required evidence of the safety of cosmetics and medical devices, and by 1949, it gained power to require safety testing of drugs and developed the first set of standards for animal testing of drugs.

Along with the evolution of our protection of food and the environment, FDA evolved in its protection of our drugs and medical devices with the development and passage of the Good Laboratory Practices Act in 1979. The GPs were put into place to govern all aspects of pharmaceutical development; they include the Good Laboratory Practices (GLPs), the Good Manufacturing Practices (GMPs), and the Good Clinical Practices (GCPs), which set standards for performing toxicology studies, manufacturing drugs and medical devices, and testing those products in humans. Those of us who worked with drug development before and after the implementation of the GPs could see a major increase in the quality of both the science and the resulting products. As an added benefit, the number of animals used in each experiment was closely controlled so that the minimum number would be used to produce meaningful results.

TOXICOLOGY TODAY

The decades since the creation of EPA and OSHA have seen numer-ous amendments to and refinements of the many laws regulating toxic chemicals. As a result, there are many sets of standards for permissible acute and chronic exposure to chemicals. Some are promulgated by public agencies and carry the full force of law. Others are merely recommendations or guidelines proposed by public or private organizations. Standards are named by the agency or act that establishes them. For example, standards for food additives and pesticide residues in foods are called tolerances. Standards for chemicals in potable water are called drinking water standards. Standards for chemical pollutants in ambient air are called national ambient air quality standards. Standards for chemicals in the air in occupational settings are called permissible exposure limits.

The same accretion of amendments and new laws has followed FDA through the years, ranging from such specifics as requiring that each batch of penicillin be examined, to total control over the approval for all medicines, including over-the-counter ones, devices, and biologics.

As in all government agencies, the enforcement of laws is political. Some administrations have been more diligent in ensuring enforcement while other administrations have been lax. Although businesses sometimes feel that compliance is costly and unnecessary, the public suffers whenever industrial noncompliance fouls the air, water, food supply, and drugs. It is important that people demand that our government continue to control industrial output of toxic waste and monitor the purity of our food and drugs.

The public should be aware that the existence of a standard does not necessarily mean that a chemical is present in the permitted quantity or indeed in any measurable quantity. The importance of this caveat became apparent many years ago when dichloro diphenyl trichloroethane (DDT) was making daily headlines. It was generally known that a tolerance of 7 ppm DDT (parts of DDT per million parts of food) was permitted in certain specific foods for which a tolerance had been requested. Logically, the public assumed that all foods, not just those for which a tolerance had been granted, contained 7 ppm DDT. In actual fact, monitoring by federal and state agencies demonstrated that DDT could be detected in only a small fraction of the permitted foods, and almost invariably in concentrations well below 7 ppm.

A major concern of many people is pesticide, antibiotic, and hormone residues that are present in foods. Although several private agencies monitor pesticide residues in foods, the results are usually in close accord with those conducted by the government. FDA conducts routine monitoring of pesticide and other chemical contaminants in more than 100 different food items, representing 11 or more different food groups, collected from retail markets throughout the United States. Adult diets have been monitored since 1960 and infant and toddler diets since 1974. The results of these studies, called "Total Diet Studies" or "Market Basket Surveys," can be found for many different chemicals at a variety of Web sites for different countries. For the United States, the FDA Web site (www.fda.gov/Food/FoodSafety/ FoodContaminantsAdulteration/ChemicalContaminants/) supplies information on various chemicals, such as dioxin and melamine, while the Web site www.fda.gov/Food/FoodSafety/ shows information for specific products, such as bottled water and baby formulas, as well as for foodborne disease. The informed consumer should check this site occasionally to see if there are any product recalls and safety alerts.

Although monitoring programs indicate that pesticide residues in fruits and vegetables grown in the United States are well within the tolerances set by EPA, there is considerable public concern that the monitoring programs themselves are inadequate. There is even more concern that imported produce is not subjected to the same scrutiny as domestic produce. Public confidence is further eroded when pesticides are misused and result in excessive residues in some food product. Other contaminants, such as melamine in pet food and milk products imported from China, have been much in the news, once again demonstrating the global nature of commerce and environmental contamination.

The public's concerns about pesticide residues in foods have resulted in the creation of a new and unofficial regulatory mechanism. A number of produce markets and supermarket chains have hired private analytical laboratories to test their produce. Such private testing programs permit markets to assure the public that there are no (or no significant) quantities of pesticides present on their fruits and vegetables. Consumers may consider the comfort provided by such an effort worth the cost. However, regulators consider the benefit to public health to be of little or no significance because routine monitoring by state and federal agencies shows no or very low levels of pesticide residues in domestic produce. Each consumer needs to decide whether this information is important to him or her or not.

The growth of the certified organic food industry is testament to the public's desire to have a safe, uncontaminated food supply. In the United States, foods that are allowed to be labeled "organic" conform to requirements set out by the National Organic Program (NOP). If plants, they cannot have been treated with nonorganic pesticides or herbicides, and if animals, they cannot have been given hormones or antibiotics. The land where these plants and animals live must be free of these specified contaminants for three years before the produce can be labeled organic. The term *organic* should not be confused with *natural*. A label of "natural" such as "all natural ingredients" has nothing to do with the certification process that allows a product to be labeled as "organic." There is no regulatory meaning to the word *natural* with the exception of meat, where the term means that no artificial ingredient (such as a coloring agent) has been added. However, for all other food categories and toiletries, manufacturers can use the term *natural* with no regulation regarding its usage.

The organic industry is one of the fastest-growing sectors of the food business. Needless to say, the consumer usually pays a higher price for fruit, vegetables, and meat that are so labeled, but many feel it is worth it to enjoy foods with none of these additives. Interestingly, because of how quickly the organic foods sector is growing, many large companies have become major players in the business. According to *The Chicago Tribune*, 2008 sales were 24 billion dollars. Companies such as Pepsi, Coca-Cola, Kraft, Nestlé, and General Mills all produce food under separate labeling for the organic market. Even companies primarily in the candy sector have entered the market with specialty companies.

The latest trend toward labeling is the term *green*, which has nothing to do with the color of the item but rather to its production in a sustainable manner. Green is not the same as organic although an item may be both.

For example, "green" fish may be those caught in a sustainable fashion, but wild-caught fish cannot be certified as organic since their food is not controlled. Farm-raised fish could be considered organic if they were fed only organic food and grown under the organic-specific conditions required for their label, but it would be hard for these fish to be labeled as "green" since the farm-raising process usually produces by-products that might not be friendly to the surrounding habitats. The recent labeling of wine as "green" and "organic" is not a contradiction; it is not impossible to meet the conditions of both using only organically grown grapes and farming the grapes under conditions of sustainability and with minimum impact to surrounding habitats. However, the "organic" label must be earned by following the NOP regulations, while the "green" label is self-administered.

After reading the discussion of organic and nonorganic in Chapter 1, the reader will realize that all food is of course organic since the chemicals that make up food contain carbon atoms. It would be impossible for humans to live on nonorganic food (although some bacteria have the ability to use sulfur-based molecules for their diet). Therefore, when speaking of "organic" food, we mean food grown under NOP conditions.

The history of the regulation of harmful chemicals in the United States over the past 100-plus years includes the response of our government to the problems that arose out of the growing use and importance of chemicals. Early on, the concern was about harmful chemicals and impurities in food and drugs. As more chemicals were developed, laws were passed to control pesticide use and pesticide residues. With continued growth, it was necessary to deal with the more complex problems of environmental pollution, waste disposal, occupational exposures, neighborhood risks, and the right to know. One of the most recent problems relating to toxic materials is their presence in waste disposal sites and their migration into groundwater. The Superfund Amendments and Reauthorization Act (SARA) was passed in 1986 to hasten the cleanup of hazardous waste sites. Almost 25 years later, relatively few of the identified sites had been remediated.

History has also shown that laws have evolved with time to meet changing needs, changing perceptions, and changing costs. Laws also have changed due to the great and rapid improvements in the sciences of toxicology and analytical chemistry. This process of change will no doubt continue for as long as required by society. In some instances, regulation involves standard-setting and enforcement; in others, it specifies manufacturing, processing, testing, or handling procedures; labeling; reporting of certain kinds of information; or record keeping.

WHAT DO TOXICOLOGISTS DO?

Toxicologists often are asked just what it is that they do. Just as there are many kinds of physicians, there are many kinds of toxicologists. Therefore, no one toxicologist will perform all the duties of toxicologists. However, here are some details about what a toxicologist might do.

In the pharmaceutical industry, toxicologists plan and conduct the in vitro (in the test tube) and in vivo (in the animal) procedures that are required for a compound or a device to be tested in humans for the first time, and then they plan and conduct the studies on that product in order for it to be registered and sold. The requirements for these studies vary somewhat by country, but with the advent of the International Commission on Harmonisation (ICH), a relatively standard battery of tests must be performed. However, since the use of the final product dictates the safety studies that need to be done, each compound or device has a specific set of safety standards that must be met. Summarizing the results, interpreting the data, and presenting them both in writing and in person to FDA are also part of a toxicologist's job. The toxicologist is an important member of the team that determines the starting dose for the first studies in humans. Additional tasks involve testing of drug impurities and compounds such as leachables (those compounds that may migrate out of glass, rubber stoppers or other bottle closure systems), the degradation products that arise in the manufacture and storage of medicinal products, and the metabolites that are produced as the body breaks down the drug in preparation for elimination.

Toxicologists in the chemical industry work on issues surrounding occupational exposures to compounds, indoor air quality, and environmental regulatory issues. Just as a pharmaceutical toxicologist must assure the safety of a drug or device, so a toxicologist in the chemical industry must ascertain the safety of chemicals to which workers, the general public, and the environment will be exposed. These chemicals include pesticides, industrial materials and by-products, combustion products, and heavy metals in our environment and in our food and water supply.

Of course, toxicologists can and do work for FDA, EPA, and other branches of federal, state, and local government. In these capacities, they act as the interpreters of rules and regulations, the writers of new regulations, and part of the team that must rule on the approval of new drugs, devices, pesticides, and other chemicals.

The forensic toxicologist, a true toxicology detective, is called in when help in needed to diagnose cases of suspected poisoning, such as when a child has eaten something unknown and the consequences of this ingestion need to be determined rapidly. These toxicologists are also involved when there are hazardous material spills resulting from truck, plane, or train accidents.

Additionally, toxicologists serve as expert witnesses for both plaintiffs and the defense in both criminal and civil litigation. Frequently, expert testimony is needed in worker's compensation cases involving exposure to chemicals; in lawsuits against doctors, hospitals, and drug companies involving overdosing or side effects of medicines; in accidents where alcohol or drug use is suspected; and in forensic cases in establishing the cause of death when poisoning is involved. Expert testimony is not as glamorous as television portrays it, since toxicologists spend most of their time wading through massive amounts of data to prepare for their 10 minutes on the witness stand.

CHAPTER 4

WHAT FACTORS INFLUENCE THE TOXIC EFFECTS OF CHEMICALS?

All effects of chemicals—beneficial, indifferent, or toxic—are dependent on a number of factors, the most important of which is the dose-time relationship: how much chemical is involved (dose) and how often during a specific period of time the exposure occurs (time). The dose-time relationship gives rise to two different types of toxicity that must be distinguished from one another: acute toxicity and chronic toxicity. The *acute toxicity of a chemical* refers to its ability to do systemic damage as a result of a one-time exposure to relatively large amounts of the chemical. Acute toxicity can occur, for example, when children are exposed to some household product or medicine left within their reach, or when some toxic substance is accidentally spilled during transportation and passersby or neighborhood residents are exposed. The exposure is sudden and often produces an emergency situation.

Chronic toxicity refers to the ability of a chemical to do systemic damage as a result of many repeated exposures to relatively low levels of the chemical over a prolonged period of time. Chronic toxicity is the concern when we evaluate the health impact of food additives, pesticide residues, drugs, or the effects of exposure to chemicals encountered by working people during the normal course of their employment. Chronic exposure to a chemical of a magnitude sufficient to produce adverse effects usually is not detected until the exposure has continued for some period of time.

ACUTE VERSUS CHRONIC TOXICITY

Why must acute and chronic toxicities be distinguished from each other? Because the symptoms produced by these two extremes of exposure usually bear no relationship to each other. Chronic toxic effects cannot be predicted

The Dose Makes the Poison: A Plain Language Guide to Toxicology, Third Edition.
By Patricia Frank and M. Alice Ottoboni
© 2011 Patricia Frank and M. Alice Ottoboni. Published 2011 by John Wiley & Sons, Inc.

from knowledge of the effects produced by acute exposures; the two extremes of exposure usually involve different organ systems or different biochemical mechanisms. That acute effects cannot be predicted from knowledge of chronic effects is also true. However, such a need never arises because knowledge of acute symptomatology of a chemical almost always precedes and surpasses that for chronic poisoning (intoxication). Acute effects of chemicals are much more readily apparent and are much more easily studied than chronic effects.

Some examples illustrate the lack of relationship between acute and chronic effects. Sublethal quantities of chlorinated solvents, such as chloroform or carbon tetrachloride, produce acute intoxication symptoms of the central nervous system—that is, excitability, dizziness, and narcosis; chronic symptoms are primarily those of liver damage. Symptoms of acute arsenic intoxication occur mainly in the gastrointestinal tract, with vomiting and profuse and painful diarrhea; chronic intoxication, however, produces skin changes and damage to the liver, peripheral nerves, and the blood-forming mechanism. Acute symptoms of lead intoxication are mainly gastrointestinal; chronic symptoms are those that result from damage to the blood-forming mechanism, nervous system, and muscular system.

Between the two extremes of acute toxicity and chronic toxicity lie a large number of combinations of quantity (how much) and frequency (how often) of exposure. Frequent exposure to relatively large amounts of a chemical can produce features of both acute and chronic symptomatologies. Adverse effects from these latter kinds of exposure are referred to as the *subacute toxicity* of a chemical.

Not only do the symptomatologies of acute and chronic intoxication by a given chemical bear little relationship to each other; neither do their relative potencies. A chemical that is highly toxic acutely is not necessarily highly toxic chronically; conversely, a chemical that is of a low order of toxicity acutely is not necessarily low in toxicity chronically.

Before proceeding further, we should explain the term LD_{50}, which commonly is used to describe acute toxicity. *LD* means "lethal dose," and the subscript 50 means that the dose was acutely lethal for 50 percent of the animals to which the chemical was administered under controlled laboratory conditions. A subscript of 0 would mean that the dose was lethal for none of the animals, and a subscript of 100 would mean that the dose was lethal for all animals tested. The units of the LD_{50} are typically mg/kg, milligrams of chemical per kilogram of body weight of the animal. A more complete discussion of the LD_{50} concept is presented in Chapter 6.

Some chemicals are acutely highly toxic but chronically in very small amounts are not toxic and perhaps even essential. Vitamin D is an example of such a chemical. Vitamin D in pure form is highly toxic acutely, with an oral LD_{50} of about 10 mg/kg, or 400,000 International Units per kilogram (IU/kg). The pesticide parathion also has an oral LD_{50} of about 10 mg/kg. If vitamin D were not exempted from the Hazardous Substances Labeling Acts by virtue of its being a food when added to milk or other edibles or a drug when sold as a vitamin supplement, in pure form it would be required to carry a poison label. Yet, despite its high acute toxicity, each one of us requires an average of 10 pg (400 IU) of vitamin D every day for good health, an amount 70,000-fold less than the LD_{50} required for a 154-pound (70-kg) person. Vitamin D deficiency results in the disease known as rickets. Severe deficiency can eventually cause death. Vitamin D can be chronically toxic when taken in daily doses several times greater than 400 IU. Daily doses of more than 2,000 IU (five 400-IU capsules per day) should be taken only under medical supervision.

People who take vitamin D supplements need not be concerned about the very high acute toxicity of the vitamin D they purchase in capsule form without a prescription. It is so tremendously diluted that each capsule contains only a recommended daily dose. For example, an acutely lethal dose of 400-IU vitamin D capsules for a young child would exceed 10,000 capsules (400,000 IU/kg body weight = 400 IU/capsule × 10 kg body weight = 10,000 capsules).

Selenium, the element with the atomic number 34, is found in various ores but also in trace amounts in soil. Selenium is a required trace mineral for mammals with some animal species showing more susceptibility than others to its toxic effects. Human vitamin supplements contain from 0 to about 70 µg of selenium. The maximum amount people can tolerate on a daily basis is 400 to 800 µg/day. Because selenium can accumulate in plants grown on selenium-rich soil, veterinary toxicity is not unknown, especially in horses and cows that graze on these lands. In a 2009 case, 21 polo ponies died in Florida after receiving an improperly mixed product that had been imported into the United States and was not approved for sale here. If the compound been mixed properly, each animal would have received an appropriate amount of selenium. Unfortunately, a fatal overdose was administered to all of the animals receiving this illegal product.

Sodium fluoride is another chemical that is acutely highly toxic but useful in trace amounts. It has an oral LD_{50} of about 35 mg/kg. Chronically, in very small amounts of 1 or 2 mg daily, it is needed for good dental health. This amount is about 1200-fold less than the LD_{50} (70 kg × 35 mg/kg = 2450/2 mg= ~1225-fold safety margin). In quantities of 3 or 4 mg per day or greater, sodium fluoride can cause mottling of tooth enamel in some young people whose permanent teeth are in the formative stages. Larger daily quantities can produce chronic fluorosis, a condition characterized by increased bone density and the formation of bone spurs.

The American Dental Association, the American Medical Association, and the Centers for Disease Control and Prevention all endorse the addition of sodium fluoride to the public water supply. People who are opposed to fluoridation of domestic water supplies take issue with the claim that the practice is safe and essential to protect the dental health of the public and cite studies that show there may be side effects to this practice. In fact, numerous and extensive epidemiologic studies of people who have lived for a lifetime in areas that have water supplies with a naturally high fluoride content show no effects attributable to fluoride, other than mottled teeth and fewer cavities. A slight increase in bone density was found in people with the highest fluoride intake, a finding not unexpected in light of the known interference by fluoride in the metabolism of calcium, an important element in bone composition.

However, further study indicated that this increased density was of no medical significance. In fact, a treatment for the severe osteoporosis associated with menopause that met with some success is oral administration of fluoride salts in quantities that approximate those ingested by people with the highest naturally occurring concentrations of fluoride in water. Natural fluoride waters contain up to 8 milligrams per liter (mg/L), which would give a daily dose of 8 to 16 mg fluoride per day. A book that discusses the controversy over fluoridation in full detail is *The Fluoride Wars: How a Modest Public Health Measure Became America's Longest-Running Political Melodrama* by R. Allan Freeze and Jay H. Lehr.

Sodium chloride—common table salt—is another chemical that is acutely toxic but essential for life in very small daily amounts. Salt has an oral LD_{50} of about 3 g/kg. If salt were not excluded from the Hazardous Substances Labeling Act because it is a food, it would be required to be labeled with a caution that it might be harmful if swallowed. Deaths have occurred among children who have been given a box of salt to play with or who have gotten into the salt shaker and ingested a large quantity of salt. A lethal dose of table salt for a one-year-old child would be about 2 tablespoons. Chronically, table salt exacerbates heart and kidney disease. In addition, excessive use of salt may be a causative or contributing factor in some cases of high blood pressure.

It has long been accepted that the critical part of the salt molecule is the sodium atom, which the normal diet supplies adequately; however, the salt molecule itself can increase blood pressure in certain sensitive individuals, so the story may be more complex than just sodium. It has been estimated that, under ordinary environmental conditions, the minimum requirement for adults is about 200 mg of sodium per day. People who work or live in very hot environments and experience excessive sweating may require a larger intake of salt to make up for the salt lost in sweat. By careful selection of low-sodium foods and by eliminating the use of salt in cooking and at the table, the intake of sodium probably can be reduced to between 1,000 and 2,000 mg per day, an amount recommended as a safe amount. The average American diet is well laced with sodium with up to 8,000 to 10,000 mg daily. A person on a low-sodium diet should recognize that many dietary components other than salt furnish sodium; monosodium glutamate, sodium citrate, and sodium bicarbonate are just a few examples.

In contrast to chemicals that are acutely toxic and chronically nontoxic, there are some that are just the reverse—acutely nontoxic and chronically toxic. Metallic mercury is one such chemical. A one-time ingestion of large amounts of metallic mercury will not cause illness or death. Thus, the frantic parent whose child chews on the thermometer and swallows its mercury contents need not fear the mercury. It will be eliminated in the feces and will cause little or no harm, because the amount of mercury absorbed from the intestinal tract is very small. But if a child swallowed the contents of a thermometer every day over a long period of time, the small amount of mercury absorbed each day could give rise to chronic mercury intoxication.

Metallic mercury must not be confused with mercury salts. The latter are very toxic both acutely and chronically. Mercury thermometers do present other risks, however. If one were dropped and broke, the mercury would be hard to clean up. It might be possible for a child to get some chronic effects from inhaling mercury vapors, which are fairly toxic and could emanate from this spill over some period of time. Many local governments encourage the removal of mercury thermometers from the house, not because of the potential for mercury poisoning of the occupants but to protect the environment, as disposal of mercury into the waste stream leads to groundwater contamination and concentrated levels in the fish that people eat. Indeed, pregnant women are advised by the Environmental Protection Agency (EPA) (www.epa.gov/waterscience/fish/advice) to limit their intake of fish to two meals per week to prevent harm to the developing fetus. Recent recommendations in the United States from some private and government sources recommend limiting fish intake by pregnant women, women of childbearing age, and children more strictly, as mercury has been found in fish from all over the world including tuna, a dietary staple in its canned form. The problem of mercury in fish is also an issue in other countries where fish is a major component of the diet and where industrial release of mercury over the years has contaminated fish.

An occupational disease associated with low-level chronic mercury poisoning is Mad Hatter's disease, named for a character in Lewis Carroll's *Alice's Adventures in Wonderland* called the Hatter and referred to as mad. You may know the Mad Hatter from the Disney cartoon of *Alice in Wonderland*. He is the zany character seen at the Tea Party. However, hat makers in England frequently went "mad" due to the use of mercury in the felting process used to make hats. The inhalation of the mercury led to neurological problems, such as garbled speech and vision problems. Mercury is not something to fool around with!

The pesticide cryolite (aluminum sodium fluoride) is another chemical that does not cause acute illness or death. The largest quantity of cryolite that can be ingested at one time will not produce illness. Cryolite passes through the intestinal tract and is eliminated without adverse effect. But if animals are fed cryolite every day in their feed, they soon become ill and die. Like mercury, cryolite is a very insoluble compound; the small amount of chemical that is absorbed from a one-time exposure is not sufficient to cause illness, but absorption of the same small amount every day, day after day, can cause chronic illness and death. As a general rule, chemicals that are insoluble or only slightly soluble tend to be nontoxic or of low acute toxicity.

A chemical that is considered completely nontoxic, acutely and chronically, is water—the fluid of life. But a few rare cases have been reported in the medical literature of both acute and chronic intoxications, some even fatal, from excessive water intake. Quantities considered excessive would be measured in gallons per day. Death from drinking excessive amounts of water occurs as a result of literally drowning the cells and tissues of the body. In 2008, there was a radio contest at KDND in California, obviously developed by people with little knowledge of toxicology, to see how much water a person could consume in a defined period of time. Unfortunately, a young woman drank almost 2 gallons of water over a short period of time and died of water intoxication; a lawsuit is pending.

Additionally, water injected intravenously can be fatal because of its ability to rupture red blood cells. This is why intravenous fluids are always given as either "normal" saline (i.e., 0.9% salt water, so-called because it approximates the amount of sodium found in plasma) or with dextrose in order to maintain the same osmolality as blood. We also know that inhaled water can be lethal as it blocks the ability of the lungs to take in oxygen and can result in drowning. In these cases, the dose as well as the route of administration makes the poison.

Most chemicals have some degree of both acute toxicity and chronic toxicity, but there is no way to estimate how toxic a chemical will be when administered chronically solely from data on the degree of acute toxicity.

Despite the general lack of correlation in severity between acute and chronic toxicities of chemicals, the effects of both acute toxicity and chronic toxicity are dose related. That is, the greater the dose, the greater the effect.

SIGNIFICANCE OF DIVIDED DOSES

The tremendous importance of the dose-time relationship in determining whether a chemical will be toxic is well illustrated by the fact that every one of us ingests many lethal doses of many chemicals, both natural and synthetic, during the course of a lifetime. A shocking thought! But consider: There is a lethal dose of caffeine in approximately 100 cups of strong coffee. There is a lethal dose of solanine in from 100 to 400 pounds of potatoes. (Solanine is a poison found in the nightshade family of which potatoes are a part.) There is a lethal dose of oxalic acid in 10 to 20 pounds of spinach or rhubarb. There is a lethal dose of ethanol in a fifth of scotch, bourbon, gin, vodka, or other hard liquor. There is a lethal dose of aspirin in 100 tablets. A listing of foods that contain potentially toxic chemicals would fill many pages. If all or even most of these were eliminated as food sources, people would suffer or die from malnutrition.

How can we ingest so many lethal doses of so many chemicals and survive? We survive because we do not take in 100 cups of coffee all at one time, or 100 pounds of potatoes, or 10 pounds of spinach, or a fifth of liquor, or 100 tablets of aspirin. We take our poisons in divided doses, not all at one sitting. Our bodies can handle small amounts of foreign chemicals, both natural and synthetic. We metabolize them or excrete them unchanged without their doing any damage. In fact, data in the toxicology literature indicate that our bodies are not just indifferent to trace quantities of foreign chemicals but that such exposures actually may be beneficial.

The concept that exposure to trace quantities of foreign chemicals actually may produce beneficial effects is unacceptable to some people. They reject evidence supporting this notion by labeling it the result of artifacts in experimental data. But rejection cannot change the fact that the phenomenon exists, and its frequency of occurrence suggests that it is not due to artifacts. This concept of *sufficient challenge*, the name given to the phenomenon by Dr. H. F. Smyth, Jr., many years ago, is explored more fully in Chapter 6.

All living organisms since the beginning of time have had to deal with exposure to numerous noxious substances. No animal on earth could survive a day, much less live to reproduce future generations, if it were not capable of handling small amounts of a wide variety of foreign chemicals. It is only when we overwhelm the natural defense mechanisms of our bodies, by taking in too much at one time or too much too often, that we get into trouble.

ROUTES OF EXPOSURE

The degree of toxicity of a chemical is dependent not only on the dose-time relationship but also on its route of exposure, the pathway by which a chemical enters the body. There are three principal routes of exposure: (1) penetration through the skin, (2) absorption through the lungs, and (3) passage across the walls of the gastrointestinal tract. These routes are labeled dermal, inhalation, and oral, respectively. Thus, by combining the dose-time and route factors, six kinds of toxicity can be identified for every chemical: acute dermal, acute inhalation, acute oral, chronic dermal, chronic inhalation, and chronic oral toxicities.

Chemicals can enter our bodies by other routes as well. A chemical may be injected into a vein (intravenously), into a muscle (intramuscularly), into the skin (intradermally), under the skin (subcutaneously), or into the peritoneal cavity (intraperitoneally). All of these routes are commonly used to study the physiology, pharmacology, and toxicology of chemicals in animal species, but human exposure by these routes is encountered only medically or in cases of substance abuse.

Dermal Exposure

A common way of contacting chemicals is by the dermal route. Fortunately, intact skin is an effective barrier against many chemicals. If a chemical cannot penetrate the skin, it cannot exert a systemic toxic effect by the dermal route. By that we mean that the rest of the body (tissues, organs, blood, etc.) is not affected since the chemical does not reach the blood circulation and distribute to organs and tissues. In this case, only the skin is involved in the response. If a chemical can penetrate the skin, its systemic toxicity depends on the degree of absorption that takes place. The greater the absorption, the greater the potential for a chemical to exert a toxic effect.

As a general rule, the majority of inorganic chemicals are not absorbed through intact skin. Organic chemicals may or may not be absorbed, depending on a number of conditions. Usually, an organic chemical in dry powder form is absorbed to a lesser degree than the same chemical in an aqueous solution, suspension, or paste. An oily solution or paste permits greater absorption than an aqueous vehicle. Certain solvents, such as dimethyl sulfoxide (commonly known as DMSO) or alcohol, enhance the dermal absorption of a wide variety of compounds. Chemicals are absorbed much more readily through damaged or abraded skin than through intact skin. Once a chemical penetrates the skin, it enters the bloodstream and is carried to all parts of the body. Exposure to chemicals by the dermal route can lead to dermal toxicities, such as irritation, rash, and acne.

The dermal route for administration of drugs can be used for compounds that are required at relatively low levels and can be placed in ointments, creams, or patches for delivery onto or through the skin. In some cases, these low levels can lead to amelioration of symptoms while staying below toxic levels. However, getting drugs through the skin is not a trivial task and sometimes requires the use of a driving force to push the drug into the body. Patches relying on low-level electricity called iontophoresis have been designed to do this with some types of chemicals.

Inhalation Exposure

Inhalation is also a common route of exposure to chemicals. Unfortunately, unlike the skin, the surface of the lungs is a poor barrier against the entry of chemicals into the body. In addition, we have a great deal more lung surface than skin surface. It has been estimated that the average adult has about 19 ft^2 (1.73 m^2) of skin and about 750 ft^2 of lung surface. (By the way, if you ever need a handy Web site to convert into and out of the metric system, go to www.metric-conversions.org.) Because of the extremely important job that the lungs do—transfer oxygen from air to blood and carbon dioxide from blood to air—that 750^2 ft area of lung surface is a very delicate, thin membrane in large part only one cell thick that separates the air in the lungs from the blood in the tissues of the lungs. This membrane allows ready passage to the bloodstream not only of oxygen but also of many other chemicals that may be present as contaminants in inhaled air. This direct access explains why people prefer to smoke marijuana rather than eat it; it gives them a greater and more rapid high. It also serves as a good example of the dependence of effect on route of exposure.

In addition to systemic damage, chemicals that pass through the lung surface may also injure its delicate, vulnerable membrane and interfere with its vital function. Chemicals such as asbestos and materials that contain crystalline silica, such as quartz dust, cause disease by damaging lung surfaces. These diseases are known as asbestosis and silicosis. Asbestos also can cause cancer in the lung and in other sites. A number of other materials damage lung surfaces, such as cotton dust and coal dust. These diseases are all grouped together under the heading of pneumoconioses. Research by scientists at Northwestern University's Feinberg School of Medicine has shown that inhalation of microparticulate matter in air pollution, such as that emitted in auto and truck exhaust, can lead to hyperclotting of the blood, which in turn leads to heart attack and stroke.

If a chemical cannot become airborne, it cannot enter the lungs and thus cannot be toxic by the inhalation route. Chemicals can become airborne in two ways: either as tiny particles composed of many molecules or atoms (dust, mists, or fumes) or as individual molecules or atoms (gases or vapors). Solids that have a low vapor pressure do not become airborne to any great extent as individual molecules, but they can be ground into very fine particles and become dispersed into the air as dusts. In general, the best size for lung absorption is 2.5 to 5 μ. Particles greater than this size do not get inhaled down to the end of the bronchioles where they can be absorbed; particles smaller than this may bounce around on the walls of the bronchioles and be exhaled. Pharmaceutical development for inhalation products must ensure that particles are the correct size, or there will be no systemic effect.

Nanoparticles are being developed as carriers for various medicinal compounds. These particles are less than 100 nanometers (nm), and many of them cannot be absorbed because of their small size. However, nanoparticles can aggregate in the bronchi to form larger particles that can then be absorbed. One of the new questions of the twenty-first century is how safe are nanoparticles, since they seem to be making their way into many consumer products, such as nanoparticulate silver–impregnated socks and T-shirts, where the antimicrobial effect of the silver prevents odors. As happens when almost any new industrial process starts, the first exposure and, therefore, the first safety concern is to those people working with in the manufacturing process. So it is with nanoparticles, workers receive higher exposure than consumers do. Because there can be many different nanoparticles (a great many compounds and elements can be produced in a nano size), the effects of these particles on worker health will be difficult and time consuming to ascertain. Studies on the effects of nanoparticles on the lungs are ongoing and will be for some time.

When particles are inhaled, the largest of the respirable particles deposit on the surfaces of the nasopharynx, or throat and, thus, do not enter the lungs. Smaller particles are breathed into the lungs and collide with the lung surfaces. As the particles become even smaller, a certain percentage of them remain airborne even in the alveolar spaces. Particles that remain airborne are breathed in and out with respiratory movements. Some of these airborne particles do, by chance, hit the bronchial or alveolar walls and remain in the lungs for a period of time. Vapors and gases in air are also breathed in and out with respiratory movements. The amount that impinges on the walls of the respiratory tract and the lung surfaces is dependent on the concentration of the material in air.

Inhaled particles that do not dissolve in the fluids coating the lung surfaces may lodge in the lungs for a long time, perhaps even permanently, or they may be swept back up the lung passages by action of the cilia that line the bronchi and bronchioles and be eliminated from the body by coughing. The latter is the more common occurrence. Cilial action is reduced in smokers who inhale, which may account in part for what is called smoker's cough. Excessive coughing may be an attempt by the body to compensate for the less effective action of the cilia. If we did not have efficient lung clearance mechanisms, those of us who live in dusty environments would soon have our lungs filled solid, a circumstance obviously incompatible with life. Chemicals that are absorbed through the surfaces of the lungs enter the bloodstream and are distributed to other parts of the body by the general circulation.

As mentioned before, large solids that enter through the respiratory passages are coughed out, coat the surface of the lungs. or are taken into the lungs where they are encapsulated by tissue to become granulomas. At autopsy, foods such as peas, corn kernels, and peanuts have been found encapsulated in the lungs. Even safety pins and other foreign bodies that lodge in lungs can stay permanently. These granulomas are benign and do not cause problems for the owner of the lungs.

Two cases of exposure by the inhalation route made the news in 2009. First, the oil company BP released excessive benzene from an Indiana oil refinery over a six-year period. The benzene was to be captured, but unfortunately it was released into the atmosphere. Although the EPA allowed BP to process 6 tons per year in this plant legally, over 16 times that amount was actually processed, overwhelming the capture system. The second case involved coal-fired power plants operated by a company at six separate sites. In this instance, the sites did not have the required pollution controls in place and were expelling high amounts of sulfur dioxide and nitrogen oxide. It was estimated by the Harvard School of Public Health that two of the sites were responsible for 2,800 asthma attacks, 550 emergency room visits, and 41 early deaths every year.

Deliberate inhalation of vapors (called *huffing*) is the cause of death or permanent brain damage in people every year. Predominately a problem among children in the 12- to 15-year-old range, this dangerous activity apparently produces a drug-induced high by using compounds that are cheap and easily obtained. In an article J. Keilman notes that inhalants can be propane, glue, rubber cement, nail polish remover, and propellants for aerosol products ("Huffing kills McHenry teen," *The Chicago Tribune*, April 19, 2010). Some of the damage is caused by the inhalants themselves, but much of the damage results from the lack of oxygen in the air that is being breathed.

Oral Exposure

The primary way that chemicals enter our bodies is by ingestion. The oral route is the principal pathway for entry of substances that are present in foods, drugs, vitamin supplements, and liquids. Chemicals that are ingested enter the body by absorption from the gastrointestinal (GI) tract. If they are not absorbed, they cannot cause systemic damage. Absorption of chemicals can occur anywhere along the digestive tract, from the mouth to the rectum, but the major site for absorption is the small intestine.

Nitroglycerine, for example, can be absorbed through the mucous membranes of the mouth, hence its medicinal administration by placement under the tongue to relieve the pains of angina. Some chemicals, such as ethyl alcohol, are absorbed from the stomach as well as the small intestines. That is why the effects of alcoholic beverages are felt so rapidly and why food in the stomach can help delay those effects.

Absorption from the stomach occurs to a much greater extent in infants than in adults. In addition, infants up to the age of about nine months have gastric juices that are less acidic, that is, more alkaline than adults. This physiologic difference is considered to be one of the reasons why nitrates, which are present in some well water, are more toxic to infants than adults. More alkaline gastric juice enhances the activity of microorganisms that reduce nitrates to their very much more toxic nitrite form.

A number of chemicals can be absorbed from the large intestine, although the primary function of this segment of the GI tract is the absorption of water. The small intestine, located between the stomach and the large intestine, is the major site for the entrance into the body of all of the absorbable substances that we ingest, which of course are primarily foods. Chemicals absorbed from the small intestine follow one of two paths. As a general rule, water-soluble chemicals go directly from the small intestine to the liver via blood in the hepatic portal vein. Fat-soluble chemicals bypass the liver by going into the lymphatic system, which empties into the bloodstream near the heart.

Ingested chemicals can also cause local toxicity to the GI tract. It is well known that aspirin in the stomach can damage the gastric mucosa. Many aspirin forms are enteric coated to reduce this stomach irritation by preventing the aspiring from dissolving until it reaches the small intestine. Nonsteroidal anti-inflammatory compounds (NSAIDs) such as ibuprofen (Advil®, Motrin®) and naproxen(Aleve®) also cause GI tract irritation "from the other side." This means that not only can their contact directly on the GI mucosa lead to damage, but their concentration in blood can also lead to GI damage by inhibiting the prostaglandins that protect the GI mucosal lining. These compounds present a serious problem, especially in older people, and can lead to damage severe enough to cause bleeding ulcers of the GI tract.

Other Routes of Exposure

Some chemicals are absorbed through the walls of the rectum; thus, certain drugs can be administered in the form of rectal suppositories. This is particularly advantageous when the oral route is not practical or possible. In addition, drugs can be administered directly into the veins (or arteries in rare cases), which is important for patients too sick to swallow or those under anesthesia during surgery and for medicines whose bioavailability (i.e., the amount that can be absorbed) after oral administration is very low. Another reason for intravenous administration is to produce an instant effect. A good example is when a patient is in cardiac arrest and epinephrine or norepinephrine is administered to start the heart. Epinephrine is available as an epi-pen for people with allergies to substances such as peanuts or bee stings. The epi-pen is really a needle that is stuck into the thigh muscle. It expels the epinephrine into the muscle from which it is then absorbed into the blood. As fast as this absorption is, it is still too slow for the treatment of cardiac arrest. Therefore, patients in a hospital setting are given these drugs directly into a vein and sometimes directly into the heart, as you can see in almost any medical drama on television. Obviously, this route, while fast and effective, is not practical outside of the physician-directed setting.

In addition to emergency treatments, many cancer chemotherapeutic agents are administered by the intravenous route. Drugs to ameliorate the effects of these agents, such as immune responses and nausea, are also given intravenously before and during the chemotherapy sessions. Other compounds commonly given intravenously are antibiotics and pain medications.

Other routes of administration used for medicines are the eye and the nose. We are all familiar with eye and nose drops, which usually deliver drugs at the site where they are needed. What is not commonly known is that many of the drugs that are given this way also end up throughout the body. Both the nose and the eye drain into the GI tract. Thus, these routes can deliver drugs to be absorbed from the intestines. Additionally, the mucosa of the nose can allow drugs to pass through to be picked up by blood in the nasal capillaries. Drugs in the eye can pass through the vitreous portion of the eye and have some small systemic availability, although this route is rarely used.

Drugs can also be administered intramuscularly, subcutaneously, and intra- or subdermally. These routes can bypass the metabolic effects of the liver and are easy enough to be used outside of a controlled medical setting. Therefore, many proteins (think insulin and growth hormone) that would be destroyed by the acidity in the stomach are administered by these routes so that their biological activity can be maintained.

Combinations

In actual practice, exposure to a chemical may or may not involve just one route. Dermal exposure can also become ingestion exposure when hands are not washed after handling or working with chemicals and before eating or smoking. Inhalation exposure can also become ingestion exposure when some of the chemical deposited on the walls of the nasopharynx is swallowed along with the secretions from that cavity, or when it is coughed up from the lungs and swallowed. Thus, a specified route of exposure is primarily, rather than solely, by that route.

INFLUENCE OF ROUTE ON TOXICITY

A few chemicals are equally toxic by all three routes of exposure. The pesticide parathion is an example. It is acutely highly toxic by skin absorption, inhalation, and ingestion. Parathion, like other organophosphate pesticides, exerts its toxic action by inhibiting cholinesterase enzymes, regardless of how it enters the body. Cholinesterase enzymes are biochemicals that our bodies use to deactivate other biochemicals known as choline esters. Acetylcholine is a choline ester that serves as a biochemical mediator of nerve impulse transmission. When cholinesterase is inhibited, acetylcholine is not deactivated, and, as a result, there is a continued stimulation of the parasympathetic nervous system. Symptoms of parasympathetic stimulation, such as excessive salivation and pinpoint pupils, then appear. The main symptoms are salivation, lacrimation (tearing), urination, and defecation, referred to by the acronym SLUD. These symptoms are classic for either organophosphate intoxication or exposure to a nerve gas such as sarin. Death can result from too great a depletion of cholinesterase activity. Exposed people (and animals) can be treated with atropine to ameliorate the symptoms.

Repeated exposures by any route to levels of parathion that produce acute symptoms may eventually result in changes in phosphate resorption by the kidneys or in electroencephalographic changes. There have been no reports of obvious clinical effects resulting from chronic exposure to parathion at concentrations that produce no change in normal cholinesterase levels. Studies with other organophosphate compounds, however, have shown behavioral and neurological changes in people chronically exposed to concentrations that do not produce clinical symptoms (see Bernard Weiss,

"Neurotoxic risks in the workplace," *Applied Occupational and Environmental Hygiene*, 5(9)(1990): 587–594).

The majority of chemicals are not equally toxic, acutely or chronically, by these routes. Vitamin D, which is highly toxic acutely by mouth, is essentially nontoxic acutely and chronically by the dermal route. Vitamin D requirements can be met by skin application, but hypervitaminosis D (vitamin D overdose) cannot be produced by this route. Nothing is known about the inhalation toxicity of vitamin D, but since it does not become airborne under ordinary conditions, the inhalation route is of no practical concern.

Metallic mercury is not considered to be acutely toxic by any route, but it is chronically toxic by ingestion. Mercury vapors are highly toxic by inhalation. A child who swallows mercury from a thermometer will not be harmed, but a child who breaks the thermometer and lets the mercury fall into the pile of the bedroom rug, or into a crack in the floor, may be in danger. Metallic mercury is a volatile liquid which is readily converted from liquid to vapor form. The degree of risk from spilled mercury depends on the concentration of mercury vapor reaching the air, which in turn depends on how much mercury is spilled, the temperature of the room, and the amount of air circulation present.

People at particular risk from mercury vapor intoxication are those who use metallic mercury in their occupations. Mercury vapor was a significant occupational hazard for personnel who worked with silver amalgam fillings in the dental profession (see M. Schneider, *Journal of the American Dental Association*, 89(1974): 1092). Thus, it is absolutely essential to observe good housekeeping procedures when working with metallic mercury in order to avoid chronic exposure to its vapors. Metallic mercury should never be stored in uncovered containers.

If mercury is spilled, it should be cleaned up immediately, no matter how small a quantity is involved. Metallic mercury is a very heavy liquid, almost 14 times heavier than water, and it has a much greater tendency to stick to itself than to anything else, with the exception of other metals such as gold or silver. Thus, when cleaning up spills, it is important to either wear gloves or to remove all rings. If you do not, you may find that your gold rings or other jewelry have turned a silvery color where they were touched by the mercury. The mercury has formed an amalgam with the gold and may cause damage to the metal.

When mercury spills, it usually spatters into many small droplets. Spilled mercury can present such a serious inhalation hazard that home vacuum cleaners should never be used to collect mercury droplets. Vacuum cleaners, unless specifically designed for hazardous waste removal, increase the air concentration of mercury by causing more rapid and complete vaporization of the droplets.

The fact that silver amalgam fillings contain metallic mercury is rediscovered periodically. The attendant publicity raises concern that people are being poisoned by their dental fillings. Numerous studies to evaluate the possibility of mercury toxicity from dental amalgams have been conducted over the more than 100 years that silver amalgam fillings have been used. Despite anecdotal reports to the contrary, there is little evidence that people with dental amalgam fillings suffer toxic effects. For a period of about a week after a silver amalgam filling is inserted (or removed), urinary excretion of mercury is slightly elevated. However, the effect is transient and well below levels consistent with mercury intoxication. Although dental amalgams are not associated with chronic toxic effects, there is the possibility that they may cause allergic reactions. Fortunately, cases of hypersensitivity to mercury are rare.

There are two basic reasons why toxicity varies with route of exposure. One relates to the quantity of chemical that gains entry into the body and the other relates to the pathway that the chemical follows in its course through the body and its metabolic fate. Ignoring the pathway and metabolism for a moment, a chemical will be most toxic by the route that permits it greatest entry. As a general rule, the lungs offer the least resistance to the entry of chemicals. Many chemicals can gain entry through the intestinal tract, although some are poorly absorbed or not absorbed by this route. The skin provides the best barrier of the three routes. For example, a chemical readily absorbed through the lung surfaces but poorly absorbed through the skin would be more toxic by inhalation than by dermal exposure. Conversely, a chemical that penetrates the skin more readily than the lungs would be more toxic by the dermal route. However, this is a rare circumstance—we are not aware of the existence of any airborne chemical that would fit in this category.

The influence of pathway through the body on the toxicity of chemicals is a very much more complex matter than degree of absorption. These influences are dependent not only on the sequence in which the chemical reaches the tissues and organ but also on the physiologic and metabolic events that occur along the way. Toxicity also depends on where target organs (those damaged by the chemical) fit in the scheme.

Ingested chemicals may be broken down by intestinal microorganisms to yield products that are more toxic, less toxic, or of the same toxicity as the parent compound. These products may or may not be absorbed from the intestinal tract in the same manner and degree as the parent compound. Thus, the oral toxicity of a chemical may differ from its dermal or inhalation toxicity merely by virtue of its passage through the intestinal tract.

The majority of compounds absorbed from the small intestines go first to the liver, via the hepatic portal vein, before entering the general circulation of the body. The liver is the great metabolic factory of the body. It plays a major role in metabolic conversions, not only of biochemicals but also of foreign chemicals. Foreign chemicals (xenobiotics) may be converted by the liver to compounds of greater toxicity, lesser toxicity, or essentially the same toxicity, or they may not be metabolized at all. After residence in the liver, any of the chemical that remains unmetabolized and its metabolic products enter the bloodstream to be circulated to all other parts of the body. Thus, assuming equal absorption by the oral, dermal, and inhalation routes, a chemical converted to a less toxic form by the liver would be less toxic by ingestion than it would by inhalation or skin absorption because it is mostly detoxified by the liver before it goes to other tissues or organs. The opposite would be true if the liver converted the chemical to a more toxic form. Some pharmaceuticals undergo almost complete metabolism by the liver before reaching the rest of the circulation. This first-pass metabolism yields the chemical almost completely nonbioavailable to the body; hence any toxicity associated with these types of compounds is due to the metabolites. Obviously, if the chemical is not available to the tissues, it cannot exert any beneficial effect either. Therefore, another route of administration must be used for these drugs.

Some chemicals that are absorbed from the small intestines bypass the liver by going into the lymphatic system rather than into the hepatic portal vein. The lymphatic system carries them back toward the heart, where they enter the bloodstream and then are carried to all other parts of the body. Chemicals that follow this liver bypass are primarily fat-soluble compounds such as fatty acids, fat-soluble vitamins, and some of the large-molecule synthetic organic compounds. These chemicals are not acted on by the liver before being distributed to other organs; thus, their toxicity is not immediately affected by passage through the liver. Eventually, however, they will be metabolized by the liver enzymes as if they had been absorbed directly into the portal blood.

Chemicals absorbed through the skin or lungs are sent directly to all other organs of the body before going to the liver, at least during their first excursion around the circuit. There may be some metabolic conversion during this passage but, if it occurs, it is usually to a much lesser degree than the conversions that occur in the liver. The molecules of a compound that are not excreted by the kidneys, skin, or lungs soon after absorption will eventually find their way to the liver. Once there, they are metabolized in the same manner as if they had gone to the liver directly, just like chemicals that are absorbed via the lymphatic system.

All chemicals that enter the body are transported in the bloodstream almost as if they had been injected intravenously. The network of blood vessels that winds through all tissues and organs is the body's delivery system, and the plasma—the noncellular component of blood—is the transport medium. Its major function is to carry nutrients to every cell and carry away waste products. All foreign chemicals that enter the body are also carried around in the bloodstream until they are disposed of by metabolism, excretion, or other factors.

METABOLISM

Before most chemicals can be eliminated from the body, they undergo the process known as *metabolism*. The investigation of the metabolic processes that occur is complex; this discussion will just scratch the surface of this area. After a chemical, whether it is a protein found in meat, a carbohydrate found in bread, or a xenobiotic found in a drug, is taken into the body, it is broken down into component parts. In the case of a protein or carbohydrate, these components are the parts of that food that are used by the body for energy and growth. In the case of a medicine or other xenobiotic, the chemical itself is used by the body to treat a disease.

Regardless of the chemical, the degradation process of metabolism begins almost instantly. It is beyond the scope of this book to discuss the metabolism of proteins, carbohydrates, and fats. It is the fate of xenobiotics, whether they are drugs or environmental pollutants, that is of interest. Many cells in the body can transform xenobiotics into other chemicals that can be more readily removed (excreted) from the body, but much metabolism takes place in the liver. There, a group of enzymes known as the liver microsomal enzyme system, or P450 enzymes (also referred to as CYPs, pronounced "sips"), convert compounds to those that are usually—but not always—less toxic than the parent compound. There are two kinds of general metabolic processes, referred to as Phase I and Phase II or Type I and Type II metabolism. Phase I metabolism involves the conversion of a compound to its metabolites. Phase II metabolism involves the coupling of either the parent compound or a metabolite to another endogenous compound, such as glucuronic acid or sulfate. Both types of metabolism result in a compound that is more readily excreted by the kidney into the urine or the liver into the bile.

Each species has its own set of P450 enzymes, but many of them are shared across species. Each CYP can be induced (geared up) or inhibited (tamped down), resulting in a complex and very hard-to-predict situation. Induction and inhibition can be influenced by the compound, the presence of other compounds and the age, sex and nutritional status of the individual as discussed below.

ROUTES OF ELIMINATION (EXCRETION)

Most foreign chemicals that enter the body eventually exit from it, either in the same form or after being metabolized to other compounds. The kidneys, GI tract, and lungs are the major organs for excretion of chemicals from the bloodstream. The kidneys might be considered the sewage treatment plant of the body. They scavenge the plasma and filter out the major portion of dissolved solids. However, they are more discriminating than sewage treatment plants in that they selectively return essential (and some not-so-essential) substances to the blood. Metabolic waste products, excess biochemicals, and the majority of water-soluble foreign compounds are eliminated from the blood in urine produced by the kidneys.

The bile, produced by the liver, serves as a vehicle for excretion of some chemicals. Bile empties into the small intestine via the bile duct, whence the compounds excreted in the bile may be absorbed back into the system or eliminated in the feces. Compounds excreted by the GI tract fall into two types: those that are excreted in their original form and those that are excreted as metabolites. Metabolites fall into two types, also: those that are excreted directly and those that are linked (conjugated) to other compounds, such as glucuronic acid or sulfur. Conjugated metabolites can be cleaved back to their original parent compound. Therefore, both unchanged compounds and conjugated metabolites can be reabsorbed. Reabsorption of foreign compounds after biliary excretion greatly complicates studies of storage and excretion of certain chemicals. For example, steroid hormones that are taken orally—frequently by athletes as performance enhancers—are recirculated by this process, which leads to high liver concentrations and greatly enhances their toxicity to this organ.

Carbon dioxide and other metabolic products that are easily converted to gas or vapor form are excreted through the lungs. Many gases, such as anesthetics, that are absorbed into the body from the lungs are also excreted in large part, or solely, by the lungs. Small organic molecules that enter the body by other routes may also be excreted through the lungs. A well-known example of the latter is ethyl alcohol. Shortly after a person imbibes an alcoholic beverage, small quantities of alcohol begin appearing in the breath. This is the basis for the breath test for drunk driving.

Other minor routes exist for excretion of chemicals. One is the skin through which small quantities of water or oil-soluble compounds leave the body in sweat or oil gland secretions, respectively.

OTHER FACTORS THAT INFLUENCE TOXICITY

In addition to the dose-time relationship and route of exposure, other factors modify toxicity, including species; sex; age; nutrition; state of health; individual sensitivity, both genetic diversity and environmental differences such as the presence of other chemicals; the phenomenon known as adaptation; and possibly light. Although the degree of influence that these factors possess may not be readily apparent, data from controlled animal experiments and from human experience demonstrate that they each is capable of exerting some degree of influence on the toxicity of at least some chemicals. The next discussion will make their influence easier to understand.

Species

The fact that species differ in their responses to the toxic properties of chemicals is of great practical significance to toxicologists. The science of toxicology depends heavily on data obtained from animal experimentation for its judgments about the toxicity of chemicals for humans. Thus, we must evaluate carefully species differences when we use data obtained from animal experiments to estimate potential human toxicity.

A large scientific literature concerns differences among species in their sensitivity to chemicals. For example, methanol, also known as wood alcohol, is highly toxic both acutely and chronically, by ingestion and inhalation, for humans and other primates. It is much less toxic for all other laboratory species, and no other known nonprimate species suffers the ocular damage and blindness that is produced in humans by methanol intoxication. Another example is the industrial chemical tri-ortho-cresyl phosphate (TOCP). Acute or chronic exposure, by all routes, to TOCP demyelinates nerve fibers in humans and chickens, but not in dogs or rats. Myelin is a biochemical that serves as a protective coating for certain kinds of nerve fibers. Demyelinization—the destruction of this protective sheath—can result in polyneuritis, the inflammation of several nerves, progressing to paralysis, which may or may not be reversible, depending on the degree and duration of intoxication.

Another chemical that also shows marked species difference is nitrobenzene, which converts hemoglobin to methemoglobin. Methemoglobin (pronounced "met hemoglobin") is an oxidized form of hemoglobin that is incapable of serving the oxygen transport function of hemoglobin. Nitrobenzene and related chemicals are more toxic acutely and chronically by all routes to humans, cats, and dogs than they are to monkeys, rats, and rabbits. The latter do not respond to nitrobenzene exposure with any significant degree of methemoglobin formation.

Acetaminophen (Tylenolreferred to as paracetamol in Europe, is a common remedy for headache, fever, and arthritis pain. Humans can take up to 4000 mg/day (~60 mg/kg) with no toxicity, although the metabolites of acetaminophen can cause liver damage in susceptible people, especially those who consume alcohol or have underlying liver damage from alcohol abuse or hepatitis C. However, one to two tablets can be fatal to domestic cats because cats metabolize acetaminophen differently from most other species.

The list of chemicals that have different toxicities for different species is extremely long and could be the subject of many volumes. The reasons for such variations are often elusive. When a mechanism is revealed, it is usually due to the fact that species handle the compound in a slightly different manner: They absorb, metabolize, or excrete it to a greater or lesser extent or at a faster or slower rate. Differences among mammalian species in metabolism of foreign chemicals are usually quantitative rather than qualitative; acetaminophen in cats is an example of a qualitative difference. Animal species possess many similar metabolic pathways for handling chemicals. The end product of metabolism depends on which of the pathways is used.

Occasionally a difference arises from some physiologic peculiarity. For example, rats are unable to vomit. Thus, when a rat ingests a toxicant, it is unable to expel the material from its stomach. This prevents the rat from eliminating the toxicant before it does damage. Dogs, in contrast, can voluntarily eject the contents of their stomachs (and frequently do!). Most dog owners are quite familiar with the heaving motions that a dog uses to induce vomiting when it has ingested some offending substance. The inability to vomit, peculiar to rodent species and probably due to different vagal nerve enervation, accounts for the apparently greater acute oral toxicity of some compounds for the rat in contrast with the dog and other nonrodent species. Such chemicals are ideally suited for use as rodenticides. The addition of an emetic to rat poison adds another layer of safety for the inadvertent ingestion of rat poison by people and pets.

Variations in toxicity of chemicals with species are of great practical importance because almost all of the knowledge of adverse effects of chemicals is obtained from toxicologic testing procedures in laboratory animals. These procedures are conducted almost entirely for the purpose of evaluating the toxicity of chemicals for humans. Thus, laboratory animals are considered to be the animal models or, more properly, surrogates for humans.

Where does the human fit in the scheme of relative toxicity? Humans do not rank consistently among animal species with regard to sensitivity to chemicals; some chemicals are more toxic for humans than for other animals, and some chemicals are less toxic. In fact, no mammalian species will always place in the same rank on listings of relative sensitivity to a series of individual chemicals, including nonhuman primates, although the chimpanzee may be closest. Because of the similarity between chimpanzees and humans, we have established a strong emotional bond with our closest species. Therefore, today chimpanzees are rarely used for research outside of government laboratories, and if they are, they may not be killed or deliberately harmed during the testing except under specific and controlled situations. After any test on a chimpanzee, the animal must be sent to a chimp "retirement home" where it will be taken care of until its natural death at a cost borne by the institution performing the test. This provision has reduced the number of chimpanzees used in biomedical research. Indeed, in 2007, the National Institutes of Health (NIH) stopped breeding chimps for research.

Proximity in the phylogenetic scheme does not assure similarity of response. Based on the evolutionary tree, it is logical that another primate, such as a monkey, would be the most appropriate animal model for humans. This is true for some chemicals but not for all, as demonstrated by the earlier references to methanol and nitrobenzene. The monkey responds as the human does to methanol intoxication, but with methemoglobin formers, such as nitrobenzene, some monkey species and humans react very differently. Thus, in the case of nitrobenzene, the dog or cat is a more appropriate animal model.

There has been a great deal of discussion in recent years concerning the extent to which data obtained from animal experimentation can be extrapolated to humans. The view that data from animal experimentation cannot be translated to humans under any circumstances has long had some proponents in the animal rights community. Such a view demonstrates a basic lack of understanding of the sciences of comparative physiology, pharmacology, and biochemistry and the principles, practices, and procedures of toxicology. If there were absolutely no relationship between humans and other mammals in their responses to chemicals, we would have no methodology for evaluating adverse effects of chemicals or, for that matter, beneficial effects of drugs, nutrients, and so forth, for the human species. The consequences of such a situation for human health and well-being would be incomprehensibly tragic. The billions of dollars spent yearly by government and industry in programs for conducting and evaluating animal toxicity tests would be a useless and unjustified waste.

Fortunately, such is not the case. There are many common threads woven through the fabric of evolution. Humans are not unique in their anatomy, physiology, or biochemistry. The similarities among mammals are far more numerous than the differences. At the molecular level, humans even show a kinship with single-celled organisms in cellular anatomy and biochemistry. For example, one of the final pathways for converting sugars, fats, and some amino acids to their end products of energy, carbon dioxide, and water is the same for almost all living organisms, whether they are plants, microorganisms, or animals. This pathway, known as the Krebs cycle or the tricarboxylic acid (TCA) cycle, was first studied in yeasts—the same kinds of organisms that are used to cause our bread to rise and to brew our beer.

Although the proposition that data obtained from animal experimentation cannot be applied directly and quantitatively to humans is flawed, there is sufficient species variation in the toxicity of chemicals to make a blind and total transfer of data from animals to humans dangerous. The opposing views—that toxicity data from animals can be transferred directly to humans and that they cannot—are both overly simplistic. The extrapolation of animal data to humans requires sound toxicologic judgment.

The first, and to this date only, legal blessing for the position that data from animals are directly applicable to humans can be found in the much-debated Delaney Clause of the 1958 Food Additives Amendments Act. This clause required that chemicals, including pesticides found in processed foods, be considered as human carcinogens if they produce cancer in any animal species at any level of exposure. It must be stressed here that the Delaney Clause applied only to processed food; it never applied to pesticides in unprocessed foods. It was never extended to any of the many other chemicals in our environment, primarily because of objections from the scientific community. With few exceptions, toxicologists rejected the concept of the Delaney Clause because it excludes the exercise of scientific judgment in the evaluation of toxicity data. Pesticide residues were removed from the Delaney Clause in the Food Quality Protection Act of 1996, although the clause still applies to food coloring agents and animal drugs in meat and poultry.

A way to determine whether the data generated in animal studies coupled with observations on human populations is to use the *precautionary principle*. This principle was designed to help governmental bodies regulate potentially harmful chemicals before they cause harm. It has been codified into law in the European Union and some other countries although not in the United States. The principle states that if a compound has a suspected risk of causing harm but the scientific community is not unanimously in agreement, then the control of that chemical should be to protect from that putative harm. If further evidence becomes available that the chemical is not harmful, then the controls can be relaxed. This principle allows regulatory bodies to protect the public while further scientific studies are conducted.

Sex

The obvious physical and physiologic differences between the sexes relate to their roles in the reproductive process as well as the effect of the male and female hormones on the rest of the body. The fact that exposure, especially occupational exposure, to certain foreign chemicals could have a profound impact on this process is generally recognized. The effectiveness of chemicals as birth control substances has been the subject of research for many decades. The subject of reproductive toxicity is discussed in Chapter 8.

Besides the reproductive process, it is known that sex does sometimes influence toxicity of chemicals in animals. Until recently, little was known about the differences between men and women in their reactions to drugs, and virtually nothing was known about the differences in reactions to environmental toxicants. In laboratory animals, some chemicals display marked sex differences in toxicity—acutely, chronically, or both. For example, male rats are about 10 times more sensitive than female rats to liver damage from chronic oral exposure to the pesticide DDT. Some organophosphate pesticides are more toxic acutely for female rats and mice than for males, whereas the reverse is true for other organophosphate pesticides.

Anatomic and physiologic differences between males and females are dictated by the sex hormones; thus, it is assumed that sex differences in toxicity are also due to hormonal influences. Indeed, there is evidence in animals to support this thesis. Sex differences in the toxicity of chemicals usually are abolished by castration or by hormone administration. In addition, chemicals that show a sex difference in adult animals often show no such difference in immature animals. However, because women have a higher percentage of body weight as fat than men, chemicals that distribute widely into fat end up with lower blood concentrations in women than in men for the same dose. This is especially important for medicines, and this fact must be factored into dosing with many drugs. The NIH has put forward guidelines to include women in clinical trials for drugs so that the sex differences in both safety and efficacy for therapeutic chemicals could be determined before these drugs were marketed.

Age

Limited information is available from human experience concerning the impact of age on the toxicity of chemicals (except drugs) in humans as compared to the data on other variables. It is known that age exerts an influence on the acute oral toxicity of some chemicals in controlled studies in laboratory animals. Newborn or infant animals may be more sensitive than adults to the adverse effects of some chemicals and less sensitive to others. For example, DDT is not acutely toxic to newborn rats. It becomes progressively more toxic as the animals mature until it finally reaches an LD_{50} of 200 to 300 mg/kg in adult rats. The opposite is true with the pesticide parathion. Although its acute oral toxicity to adult rats is very high, it is even more toxic to newborn animals. Data from accidental poisonings with parathion indicate that the same is true for the human species. The quantities of parathion that have caused illness or death among children are considerably less than predicted from calculations based on quantities that have produced the same effects in adults. Boric acid is another chemical that seems to be much more acutely toxic orally to human infants and toddlers than to adults.

Differences between young and adult animals in their responses to certain chemicals are considered to be due in part to differences in the activity of the enzyme systems that metabolize foreign chemicals (see "Metabolism" section in this chapter). These systems are underdeveloped in immature animals. Thus, a chemical converted to a less toxic metabolite by liver microsomal enzymes theoretically would be less toxic to adults than to infants. The fully developed liver microsomal enzyme systems of the adult would metabolize the toxic chemical more quickly and effectively than the immature infant systems would. Conversely, a chemical converted to a more toxic form by the liver microsomal enzymes would be more toxic to adults than to infants.

It appears logical that children could be more affected by exposure to chemicals than adults. Infants and young children do consume considerably more food per unit body weight than adults. During the early years, the need for energy and nutrients from food for body growth and development is greater than at any other period of life. As the child develops, physical activity increases greatly and adds to the child's energy demands. Thus, if a child consumes relatively more of a food than an adult and that food contains a pesticide residue or any other chemical, the relative intake is greater in the child. This coupled with the lessened metabolic activity of the liver can lead to more toxicity in children.

A critical point is how the exposure of infants and children to compounds relates to levels of those compounds considered safe, such as

Acceptable Daily Intakes (ADIs) set by the World Health Organization. (EPA sometimes uses the term *RID* or *RFD*, meaning Reference Dose, to express the same concept as ADI.) Data from total diet studies conducted by EPA and the Food and Drug Administration (FDA) (see Chapter 3) show that the diets of infants and toddlers contain quantities of pesticides only about one hundredth of their respective ADIs. ADIs, in turn, are set at one hundredth or less of levels considered safe.

The majority of pesticides have not been shown to be carcinogenic. Among those that have caused cancers in animal studies, the majority do not cause cancer unless they are converted in the body to another chemical that is a carcinogen (see Chapter 7). Many pesticides are metabolized as foreign chemicals by the liver microsomal enzymes (P450s) mentioned earlier. Since immature enzyme systems do not metabolize foreign chemicals efficiently, and hence cannot convert foreign chemicals to a carcinogenic form as readily, children may be more protected than adults from any carcinogenic properties of pesticides. Probably the greatest carcinogenic risk for growing organisms is exposure to ionizing radiation, which can contact DNA molecules directly. For the majority of humans, such radiation exposure is from natural sources.

Having said all this, exposure to pesticides in infants and children is of concern for reasons other than cancer. In 1996, EPA toughened the standards for testing of pesticides to determine their effects on the nervous system and endocrine systems since the toxic effects of pesticides may be subtle and hard to detect (EPA fact sheet, *Protecting Children from Pesticides*, www.epa.gov/ pesticides/factsheets/kidpesticide.htm).

In 1999, FDA implemented the pediatric initiative in order to generate more applicable data on the effect of drugs on children. This initiative requires pharmaceutical companies to study medicines in infant and juvenile populations. In 2002, the Best Pharmaceuticals for Children Act was enacted, providing more funding and a bigger commitment on the part of the government to ensure that new and old drugs were adequately tested in children. Of course, the testing of drugs in children raises ethical questions about whether they should be subjected to experimental medicines without their "informed consent," as is required of adults, and whether adults can make informed consent for children. This type of social issue demonstrates once again that research is not performed in a vacuum and that the positives derived from one stance may be counterbalanced by negatives. We address this dichotomy in more detail when we discuss benefits and risks in Chapter 11.

Older people may be more susceptible to the toxic effects of chemicals than younger adults. Examples of this phenomenon are especially prevalent among pharmaceuticals. Again, FDA is requiring more studies in the geriatric population to try to determine if side effects are more severe or more common. Some reasons for more toxicity are due to the deterioration of organ systems with age, changes in the metabolizing systems that may lead to higher plasma concentrations of a drug, and the presence of multiple drugs in the body at the same time. For example, aspirinlike drug (NSAID) administration has been shown to lead to a greater incidence of GI upset and ulcers in older people, primarily due to a lessening of the natural protective elements found in the gut.

Nutrition

Diet plays an important role in the toxicity of chemicals in laboratory animals. It is assumed that diet also modifies the toxicity of chemicals for humans. As a general rule, diets adequate in proteins and vitamins protect against some of the toxic effects of chemicals. The simplicity of this rule belies the great complexity of nutrition-toxicity interactions. This complexity is further compounded by data showing that laboratory animals fed diets nutritionally adequate but restricted in quantity, without the addition of any added foreign chemicals, develop significantly fewer tumors than animals fed all that they want to eat of exactly the same diet. Calorie-restricted diets have repeatedly shown an anticancer effect. A relatively small group of people are following such diets, although it is quite a challenge for those of us who both like to eat and do not like to feel hungry. The complex subject of nutrition is beyond the scope of this book.

State of Health

A person's individual response to toxicants is influenced by his or her physical and emotional health. For example, certain physical conditions, such as liver disease or lung disease, enhance the toxic effects of chemicals that cause liver or lung damage, respectively. Medical experience has shown that table salt and possibly other sodium-containing salts are chronically more toxic to people with heart or kidney disease than they are to people with normally functioning hearts and kidneys. Very few controlled animal experiments reported in the scientific literature relate physical health to chemical toxicity. However, scientists who conduct animal toxicity studies have long recognized that laboratory animals must be in good health in order to be valid experimental tools. Sick or maltreated animals do not yield reliable results. However, the effect of drugs on sick people may be very different from the effect of an equivalent dose of drug on healthy animals.

The impact of emotional health on toxicity of chemicals is an even more difficult subject to study than the influence of physical health. Convincing data in the medical and dental literature indicate that various emotional stresses adversely affect physical health. Thus, there is an intuitive sense that they also must exert some influence on toxicity, even if only indirectly. Many articles on the influence of stress on such medical conditions as cancer and allergies have appeared in the popular press in recent years.

Biochemical Individuality (Genetic Diversity)

The reactions of individuals vary with all of the preceding factors, but, in addition, much evidence demonstrates that a person's biochemical makeup can also modify the toxicity of chemicals. This factor is referred to as individual susceptibility, biochemical individuality, or genetic diversity. The most obvious expression of differences among individuals in their reactions to chemicals is seen in their acute responses. There is no such thing as a single dose that will have exactly the same degree of effect in all individuals. For some people, one aspirin tablet will cure a headache, whereas it may take two or even three tablets to be effective for others. Further, consider the differences among people in their tolerance or lack thereof to alcohol.

Individual susceptibility to chemicals occurs in all species. That is why acute lethal doses are expressed as averages or means rather than as absolute values. For example, 10 mg/kg of parathion will not kill every rat. It will kill approximately half of a group of rats, but the remaining half will require a larger dose for a lethal effect, and among the half that died, most would have succumbed to a smaller dose.

It is known that some differences in susceptibilities are based on differences in genetic makeup (one's *genome*). A number of well-defined and documented genetic traits render their possessors more susceptible than the general population to the adverse effects of chemicals and physical agents. For example, people who have red blood cells that are more fragile than normal due to a certain enzyme deficiency are more susceptible to chemicals that cause hemolysis. People who have a genetic deficiency of a certain DNA-repair process suffer from xeroderma pigmentosa, a disease that renders them tremendously more susceptible to the carcinogenic effects of ultraviolet radiation on the skin. People with albinism, and fair-skinned people in general, are more sensitive to the damaging effects of sunlight.

Anyone interested in the subject of genetic variations in susceptibility to specific chemicals and physical agents should consult *Biochemical Individuality*

(1998) by Roger J. Williams. Williams is generally credited with the recognition and development of the concept of biochemical individuality.

Genetic variations in susceptibility to chemicals are often called genetic defects—a misnomer since some genetic variation can be protective rather than damaging. For example, people who have a lowered ability to produce the enzyme AHH (aryl hydrocarbon hydroxylase) do not as readily metabolize polycyclic aromatic hydrocarbons, such as benzo[a]pyrene, by hydroxylation. Since scientific data indicate that this compound must be hydroxylated in order to become the potent carcinogen that it is known to be, the inability to produce AHH is protective. Genetic variation may be responsible, at least in part, for the fact that not all people who are heavy smokers of long duration develop lung cancer. The lack of AHH could be protective against the carcinogenic effects of the polycyclic aromatic hydrocarbons in tobacco smoke.

When we observe people who have hypersusceptibilities that cannot be explained by current knowledge, it becomes obvious there is still a great deal more to be learned about individual reactions to chemicals. Some of these cases are so implausible that the victims tend to be dismissed as eccentrics. The people who have contacted one of this volume's authors (MAO) to discuss their extreme hypersusceptibilities to a variety of environmental chemicals, however, have been intelligent, often highly educated, and reasonable people who were not given to hysterics. Fortunately, severe hypersusceptibilities are relatively rare, but for the individuals so afflicted their illnesses are personally tragic.

For some, the problem may be allergic in nature, but for others, there appears to be no answer at the present time. However, there is a ray of hope. The exciting advances in the understanding of immunologic tolerances and intolerances to biochemicals within one's own body and to foreign chemicals are persuasive evidence that ultimately the solution to many cases of hypersusceptibility will come from the science of immunology.

With the advances in our understanding of the human genome, the genes responsible for many biochemical responses are and will continue to be identified, and we will increase our understanding of biochemical variability. Already the impact of some genes on disease states, such as Parkinson's disease and diabetes, has become clearer due to the genetic sequencing of various human populations.

Presence of Other Chemicals

The toxicity of chemicals can also be modified by the presence of other chemicals. In some cases, the toxicity may increase over the sum of that seen for each individual chemical (synergism); in others, toxicity may decrease (antagonism). When chemicals act synergistically, the toxic effect observed is greater than would be predicted from data for the individual chemicals; that is, the effects are more than simply additive. Synergism might be likened to 2 plus 2 equaling 5 (or any other number greater than 4). When chemicals act antagonistically, the toxic effect of the combination is less than what would be predicted from the individual toxicities (2 plus 2 equaling 3 or less).

The best-understood mechanism whereby chemicals synergize or antagonize one another is interference in the metabolism of one chemical by the other. Thus, synergism and antagonism may be likened to two sides of a coin. If metabolism converts a chemical to a more toxic form, inhibition of metabolism by another chemical will prevent that conversion, and the toxic effect will be less than predicted (antagonism). If metabolism converts a chemical to a less toxic breakdown product, inhibition of metabolism by another chemical will prevent detoxification, and the resulting toxic effect will be greater than predicted (synergism). If a chemical increases rather than inhibits the metabolism of another chemical, these results would be reversed. The chemical would increase (synergize) the toxic effects of a chemical converted to a more toxic form and decrease (antagonize) the effects of a chemical converted to a less toxic form.

Frequently, chemical interactions result from two chemicals competing for the same P450 metabolizing enzymes. Such a situation is exemplified by the restriction on eating grapefruit or drinking grapefruit juice while taking over 50 different medications, such as the statin cholesterol-lowering drugs (Lipitor®, Zocor®, etc.) and heart medicines known as calcium channel blockers (Procardia®, Norvasc®, etc.) as well as other drugs. In these cases, compounds in grapefruit juice block metabolic enzymes in the small intestine that limit absorption of these drugs; thus, in the presences of grapefruit juice, more drug is absorbed, resulting in higher concentrations of the medications in the body. This can lead not only to greater possible efficacy for the drug—a desirable effect—but also to the more troublesome increase in toxic side effects. For drugs such as statins and calcium channel blockers, increased blood levels can lead to serious side effects. However, for a few other classes of drugs where it is difficult to achieve high enough therapeutic concentrations of medication, concomitant grapefruit juice can be used to advantage to increase drug levels to an effective, but still safe, concentration in the blood. One such drug is sirolimus (rapamycin) which is used as an anticancer therapy. This drug is very expensive (more than $10,000/year) and is given long term to prevent the formation of new blood vessels in tumors, so the ability to use less drug with the same effect would be a real money saver. According to a study undertaken by doctors at the University of Chicago Pritzker School of Medicine, fresh-squeezed grapefruit juice is the key to success as it renders a lower dose as effective as the standard, higher dose.

Considerable data exist in the scientific literature on synergism and antagonism of acute effects between individual chemicals and classes of chemicals, but almost nothing relates to chronic effects. An exception is the well-established fact that smoking synergizes the carcinogenic properties of asbestos. Based on studies of asbestos workers, the risk of lung cancer in a smoker exposed to asbestos is 20 to 30 times greater than a nonsmoker exposed to the same asbestos concentrations.

The lack of information about chronic synergistic effects is not because chronic interactions do not occur or because they are not important but rather because the study of such interactions is difficult and costly, both in time and money. Chapter 5 describing methods for conducting toxicology studies will help to clarify why this is so. The problems involved with the chronic toxicity testing of individual chemicals are greatly compounded when consideration of synergism and antagonism by other chemicals is added to the protocol. There are so many thousands of chemicals, both natural and synthetic, to which some human populations have a significant exposure that just deciding which combinations of chemicals should be tested, and in what order, becomes a Herculean task. However, the difficulties do not dissuade toxicologists from interest in the subject of chronic interactions among chemicals or from attempts to design experimental methods to overcome them.

Adaptation

Adaptation is a term applied to the process whereby exposure to subtoxic doses of a chemical renders a person tolerant to subsequent doses of the chemical in quantities that would be harmful to nonadapted individuals. Perhaps the most dramatic example of mass adaptation is that of the fabled arsenic eaters of Styria. People of this mountainous region in central Europe ate small quantities of arsenic trioxide, found naturally in the area, once or twice a week for health purposes. They finally accustomed themselves to doses of as much as 400 mg of arsenic trioxide a day, a quantity that would cause serious illness or death in ordinary people. The arsenic eaters were reputed to have had longer-than-average life spans.

Adaptation can occur with many chemicals. Two examples most familiar to the general public are alcohol and nicotine. Many people who can handle a three-martini lunch can remember when a one-martini lunch produced the same glow. And probably few smokers can forget the nausea produced by that first covert cigarette. The phenomenon of adaptation was referred to as *habituation* in the early medical literature. It was considered to be due to changes in degree of absorption of the chemical involved. Today, with the increased understanding of biochemical mechanisms, it is recognized that many cases of adaptation are the result of responses by enzymes that process the chemical in question.

The discovery of adaptive enzymes—also known as inducible enzymes—is a fascinating story that has its beginnings in the early research into the metabolic fate of simple nutrients in single-celled organisms. It was long known that if yeast cells were placed in a medium containing glucose, the monosaccharide of table sugar, they started multiplying immediately. It was also observed that if yeast cells were placed in a medium containing galactose, a monosaccharide of milk sugar, they stayed quiescent for a while, then started multiplying slowly, then faster and faster, until they finally achieved the same rate of increase as if they had been placed in glucose solution to begin with. Further, when the yeast cells that had been forced to grow on galactose were harvested and then returned to a medium containing galactose, they did not show the lag time that they had previously but rather started multiplying immediately. The necessary enzymes were already available as a result of their prior exposure to galactose.

This behavior of yeast cells, combined with many other similar observations in yeast and other microorganisms, led to the theory that the potential to produce certain enzymes is present in all cells but that the actual enzymes are not produced in any quantity until they are needed. Thus, yeast cells placed in a galactose medium do not grow immediately because the enzymes needed to metabolize galactose are not present to any extent. But the ability to gear up the enzyme production is present, so that when yeast cells are offered galactose as their food source, they slowly form the enzymes required to metabolize it. These enzymes are induced by the presence of the chemical they metabolize in the media in which they grow. How very efficient! Why expend the energy required to produce an enzyme if it is not needed, or before it is needed?

Since the discovery of inducible enzymes, it has become recognized that all of us, within the cells of our bodies, contain many inducible enzymes. In fact, some biochemists believe that all enzymes are inducible and that those that are always present in our cells are those for which the substrates, such as the common nutrients, are always present. Inducible enzymes explain why we can develop a certain amount of tolerance to a number of foreign chemicals. The ability to produce these enzymes, accompanied by trace quantities of the actual enzymes themselves, is present and waiting in our cells for the substrate to come along to be acted on. As the concentration of substrate increases, so does the quantity of enzyme, up to a level dictated by cellular biochemistry.

The process of adaptation by means of inducible enzymes is not available for all foreign chemicals, nor can it protect a person totally against ever increasing amounts of toxicant. Even inducible enzymes can be overwhelmed by too much toxin too often.

Light

Light, artificial or natural, does not usually influence the toxicity of chemicals. A possible exception is the effect of light on the metabolism of bilirubin in the blood of newborn babies. Bilirubin, a yellow-colored, biochemical waste product, is discharged into the bloodstream when excess red blood cells are destroyed by the body shortly after birth. This is a normal process. However, unless this bilirubin is promptly metabolized and excreted, it can build up in the blood to the point where it causes a serious disease known as neonatal jaundice (hyperbilirubinemia). This common disease of newborns used to be difficult to treat. Today, because doctors know that light greatly enhances the metabolism of bilirubin in newborn babies, they are exposed to light, which is a safe and effective treatment for neonatal jaundice.

It has long been known that light exerts a profound influence on physiological responses in animals, including control of reproductive cycles in some species. In the early twentieth century, it was discovered that exposure to sunlight could prevent the dreaded bone disease rickets. The discovery of the role of sunlight in vitamin D synthesis in human skin followed soon after. Scientific interest in the influence of light on human physiology and pathology increased in succeeding decades. As a result, the observation that some people suffered increased periods of depression during fall and winter months led to the recognition of a disease entity now known as seasonal affective disorder (SAD). The symptoms of SAD are fatigue, sadness, excessive sleepiness, craving of sweets, and weight gain. Proof of the role of light is the successful treatment of SAD with light therapy. Exposure to bright light for an additional five to six hours a day abolishes the symptoms of SAD.

The fact that light is physiologically active in humans and animals justifies its inclusion in this chapter. Photosensitization reactions described in Chapter 2 may be a manifestation of an influence of light on the toxicity of chemicals.

CHAPTER 5

HOW IS TOXICOLOGY STUDIED?

The purpose of this chapter is to give the reader an introduction to the kinds of experiments employed to study the adverse effects of chemicals, the limitations of the methods and analytical procedures involved, and the meaning of some of the terms used to describe effects or quantities of exposure. It is not intended to serve as a cookbook for the conduct of toxicity tests. It is beyond the scope of this book to provide more than a brief summary of methods of toxicity testing because they are too many and varied. In some cases they are ingenious. Details of testing protocols, both animal and nonanimal, can be found in toxicology and pharmacology texts and journals. In addition, regulatory agencies, such as the Food and Drug Administration (FDA) and Environmental Protection Agency (EPA), have formalized the testing requirements and experimental protocols for chemicals that come under their regulatory purview. This information can be obtained from the Web sites of the respective agencies under "guidances" (www.fda.gov; www.epa.gov). International harmonization of the testing performed for certain types of compounds has been ongoing for many years. The International Committee for Harmonisation also has a Web site that has guidances that can be downloaded (www.ICH.org).

The key to relevant toxicity testing in animals is the selection of a species that handles the chemical under study in the same manner as humans do with regard to absorption, metabolism, excretion, and so on. Before such a model can be selected, extensive study into the physical and biochemical properties of the chemical may have to be undertaken. Since there is no one species that handles all chemicals in the same manner as humans, there is no one animal model that can be used to study the toxicity of all chemicals for humans. The identification and characterization of animal and nonanimal models for research have given rise to an important new science based on comparative toxicology.

The Dose Makes the Poison: A Plain Language Guide to Toxicology, Third Edition.
By Patricia Frank and M. Alice Ottoboni
© 2011 Patricia Frank and M. Alice Ottoboni. Published 2011 by John Wiley & Sons, Inc.

Among the earliest subjects of toxicity testing were slaves who served as tasters for emperors, kings, and other nobles who feared that some unknown enemy might have slipped poison into their food or drink. Modern medical science brought the need for the development of surgical techniques and the study of the efficacy of medications and the use of animals became common in the research laboratory. Undoubtedly, the first experimental animals in modern times were stray dogs and cats. As the need for pharmacological research increased, so did the need for large numbers of experimental animals. Rats and mice were ideal subjects; they were prolific and small. Thus, they could be produced in large numbers and, because of their modest space requirements, could be housed inexpensively. Wild mice, rats, and other rodents that were trapped many decades ago and bred in captivity became the ancestors of today's commercially produced laboratory animals, animals of high quality and known lineage, thus minimizing genetic variability. While inbreeding to produce strains with similar genetic backgrounds allows for reproducible testing, it leaves out the effect of genetic variability on toxicity.

EXPERIMENTAL METHODS

The chemicals subject to routine toxicity testing are mainly those new chemicals coming into general commercial use. These include such chemicals as pesticides, drugs, food additives, industrial chemicals, and household products. In addition, older chemicals in any category that have come under suspicion since their commercial introduction for causing cancer or other adverse health effects are subjected to retesting. Also, continuing research on vitamins, human hormones, enzymes, and proteins and vaccines to prevent and treat diseases generates toxicity testing for these types of compounds. Animal testing is not performed solely on chemicals but also on devices, such as cardiac stents, dialysis tubing, and others.

Since the carcinogenicity of chemicals is of great concern, it should be pointed out that only a small fraction of the many thousands of chemicals that have been subjected to toxicity testing or retesting as a result of regulation by EPA or FDA have been found to be carcinogenic in animals. Because of the patterns of home use of chemicals and the increasing federal regulation of the use of cancer-causing materials, the percentage of potentially carcinogenic chemicals that find their way into the home setting would be an even smaller fraction. Positive carcinogenic findings for medical products usually keep these compounds off the market unless they are for the treatment of a very serious disease, such as cancer itself.

Acute Toxicity

The methods used to study the acute toxicity of chemicals are relatively easy to perform and relatively inexpensive. (*Acute testing* refers to administration of the chemical once; *subacute testing* refers to administration for up to one month; *subchronic testing* refers to up to three months; and chronic *testing* refers to longer than 3 months.) The purpose of acute testing is to show what might happen in case of accidental or deliberate exposure to large amounts of chemicals or products. Probably the majority of accidental acute poisonings occur in children, but adults also can be exposed to acutely harmful amounts of chemicals in home or work situations. The general public is at particular risk when chemicals are released in large quantities during transportation accidents.

The classic test for the study of acute toxicity is known as the LD_{50} test. *LD* means "lethal dose," and the subscript 50 refers to the percent of the test animals for which the test chemical is lethal. This test was the most commonly performed acute toxicity test, and it consumed many hundreds of thousands of animals yearly worldwide. Most synthetic chemicals available for commercial or home use before 2000 have been subjected to an LD_{50} study.

The classic LD_{50} experiment, which is described next, is by and large no longer required by U.S. regulatory bodies for chemicals they regulate since other, animal-sparing methods give similar results; however, because so much LD_{50} data exist, the method used to determine LD_{50} values is presented here. Data about acute toxicity are still required for many chemicals, although FDA is no longer requiring specific acute toxicity studies for new drugs since the information generated in longer-term studies generally gives sufficient acute data as well.

In the absence of acceptable no-animal methods, the only alternative is to design studies that use fewer animals than the classic LD_{50} method. This was the approach taken by EPA in 1988. In a revised policy on acute toxicity testing, EPA offered several methods that would reduce the number of animals used without compromising public safety. For example, the standard LD_{50} test might employ up to 100 animals since the usual number of animals per group would be from 3 to 5 for each sex with many groups established. The up-and-down method allows 1 animal to be dosed and, based on the results of that dose administration, the next dose to a single animal would be either higher or lower. Thus, as few as 5 to 10 animals would be used in this type of acute toxicity testing. Because the results of the classic LD_{50} test are not very accurate and extrapolation of the results to humans is inaccurate, using a test that has somewhat less accuracy is of no practical significance. Currently, both the U.S. EPA and FDA discourage the use of the LD_{50} test. For a full discussion of various other alternate methods to the standard LD_{50} test, see A. W. Hayes, *Principles and Methods of Toxicology*, 4th ed. (2001).

The test animals for classic LD_{50} experiments are usually rats and mice. Preliminary studies provide a general idea of whether the chemical is of high, moderate, or low acute toxicity. This information supports decisions regarding the doses to be administered in the LD_{50} test. The test animals are divided into four or five dosage groups of several animals each. Each animal in the lowest dosage group is administered an amount of the substance that preliminary tests indicate will cause no deaths. Each animal in the highest group is given a quantity of the substance calculated to be lethal to most of the animals. The other groups are given quantities intermediate between the two extremes. The animals are observed closely for a 14-day period. All adverse reactions and deaths are recorded during the observation period. Then the data are plotted on a graph to yield what is known as the dose-mortality curve. The quantity of chemical that falls on the point of the curve corresponding to 50 percent mortality is the value taken for the LD_{50}.

LD_{50} values can be determined for oral, intravenous, and dermal routes of exposure. For an oral toxicity test, the chemical is administered by stomach tube. Environmental chemicals with oral LD_{50} values of 50 mg/kg (milligrams per kilogram) or less are classed as highly toxic (poisonous). Chemicals with oral LD_{50} values of 50 to 500 mg/kg are considered to be moderately toxic. The toxic range extends up to 5 g/kg for environmental chemicals and 2 g/kg for medicinal chemicals. Chemicals with oral LD_{50} values greater than those are classed as outside the orally toxic range, or nontoxic by default. Since any chemical is capable of causing illness under some set of circumstances, toxicologists prefer to consider the latter group as having relatively low toxicity rather than no toxicity. Intravenous LD_{50} values are determined the same way as oral LD_{50} values, but the compound of interest must be soluble in a biocompatible vehicle (e.g., saline or certain buffers) and is injected into a vein instead of placed in the stomach.

Dermal LD_{50} values are determined by placing a weighed quantity of chemical in continuous contact with an area of bare skin that is approximately 10 percent of the animal's total body surface. The test may be conducted on both intact skin and abraded skin. Care must be taken that the animals do not lick the treated area since this will confuse the dermal toxicity with toxicity associated with ingestion of the material. Rabbits usually are used in dermal toxicity studies because of their larger size and, hence, their larger areas of skin. The period of continuous contact is 24 hours. Chemicals with dermal LD_{50} values below 200 mg/kg are placed in the highly toxic category. Those with dermal LD_{50} values between 200 mg/kg and 2 g/kg are classed as toxic. Dermal LD_{50} values greater than 2 g/kg are outside the toxic range.

The measure of acute toxicity by inhalation is termed the LC_{50}, lethal concentration for 50 percent of the animals tested. LC_{50} values are determined by exposing several groups of animals, usually rats or mice, each to a different air concentration of a chemical, for a 1-hour period followed by a 14-day observation period. Alternatively, the same concentration can be used but exposure periods can vary. At the end of the observation period, the data are plotted to give an air concentration–mortality curve similar to that for oral, intravenous, or dermal toxicity. The air concentration that corresponds to 50 percent mortality is taken as the LC_{50}. Chemicals with an LC_{50} of up to 20,000 parts of gas or vapor per million parts (ppm) of air or 200 mg dust per liter of air (mg/L) are classed as toxic by inhalation. Those with LC_{50} values at or below 200 ppm gas or vapor or 2 mg/L dust are in the highly toxic (poison) category. Chemicals with LC_{50} values greater than 20,000 ppm gas or vapor or 200 mg/L dust are considered nontoxic.

Irritant and Corrosive Effects

Testing for irritant or corrosive properties is standard procedure for substances regulated by various federal agencies. Such testing is important to protect people who contact irritant or corrosive chemicals in their occupations. It is even more important for the protection of members of the general public who bring irritant or corrosive products into their homes, particularly homes with small children. The general public usually has less knowledge about how to protect itself and less access to protective equipment than do workers.

Oral and dermal LD_{50} studies provide information about irritancy and corrosiveness of chemicals as well as their toxicities. In fact, the majority of corrosive chemicals, such as the drain cleaners found under many kitchen sinks, have LD_{50} values that place them in the dangerous range. These chemicals are not highly toxic, but their corrosiveness makes them lethal in very small amounts as they chemically destroy the tissues they contact.

Many products composed of chemicals that are known to be of relatively low toxicity may not require LD_{50} studies, but some investigation of the irritancy properties of these chemicals must be conducted. Most desirable are tests that give rapid results. Many such tests are available today, animal, nonanimal, and human.

The classic techniques for studying irritation and corrosiveness to skin and eyes were developed many decades ago. They bear the name of J. H.

Draize, who was chief of the Skin Toxicity Branch of FDA at the time the tests were developed. The skin tests follow the dermal LD_{50} protocol, except that the holding period is 7 days rather than 14. The degree of irritation is determined by rating the degree of redness (erythema), swelling (edema), and/or blistering according to a standardized score chart. The use of human volunteers as subjects for studies of skin irritant properties of substances whose toxicological properties are quite well understood, such as soaps and detergents, is relatively common and probably the most informative and useful method of study of mild irritants.

No area of animal testing has given rise to more public outcry than the Draize test for ocular reactions. In this test, albino rabbits are used. The test material is dropped into one eye of each rabbit, with the other eye serving as a control. In one group, the treated eye is not washed after instillation of the test material; in the other group, the treated eyes are washed. After 24, 48, and 72 hours, and at 4 and 7 days, the results are rated according to a standardized score chart. Of the many attempts to substitute in vitro methods for this test, none has yet been totally accepted.

Probably no industry has a greater need for tests of irritation or damage to skin and eyes than the cosmetic industry, which, in collaboration with drug and chemical companies, has been testing techniques that would eliminate the need for live animals. Materials such as tissue cultures, chicken egg membranes, and vegetable protein films have shown promise as substitutes for chemicals with severe corrosive effects. However, the research and validation demonstrating that any one (or more) are as predictive of damage as the Draize test are not yet forthcoming. At the present time, there is no substitute for an animal eye to detect chemicals that have only mild to moderate eye irritant effects. In May 2009 the European Centre for the Validation of Alternative Methods (ECVAM) reported that after review of multiple in vitro methods to replace the Draize ocular irritancy test, it needed more time before making a recommendation, but two methods did appeared useful in limited cases for cosmetics testing if care was taken in interpreting the data. These methods could not be used for ocular lens cleaning agents, however, so the search for alternate methods continues.

Sensitization and Photosensitization

An allergic reaction to some agent is a common condition for which people seek medical attention. Allergies may be fatal, especially to some antigens, such as peanuts and bee stings, but they rarely are. However, such allergies are often distressing and in extreme cases debilitating. People are so heterogeneous in their susceptibilities to sensitization that it is a rare chemical that has not caused an allergy in some individual. Thus, the major goal of testing a chemical for sensitizing properties is not to discover whether it can be a sensitizer but rather the strength of its ability to sensitize—that is, the percentage of an exposed population it will affect. Because of product liability, it is understandable that information about the ability of a chemical to cause an allergic reaction is probably as important to the cosmetic industry as information about the chemical's irritant or corrosive properties.

Guinea pigs are the animals mostly used for sensitization studies because they are extremely susceptible to a wide variety of chemical sensitizers. In one of the common protocols, small quantities of the test chemical are injected within the layers of the skin over a small area of the back or sides, one injection every other day, until a total of 10 sensitizing injections have been made. After a rest period of two weeks, a challenge injection, smaller in quantity than the sensitizing injections, is administered at a site just below the area of the sensitizing injections. Twenty-four hours later, reactions are rated according to a standardized score chart.

Obviously, the ideal animal models for sensitization experiments are humans. Before humans are used, some information must be available about how severe reactions might be. Exposure to severe sensitizers could produce long-term adverse effects or even death in human subjects. The guinea pig test is a valuable tool for screening chemicals for the strength of their sensitizing properties prior to human exposure.

Photosensitization studies are conducted similarly to sensitization studies. Since photosensitization may occur after oral as well as dermal exposure, the test chemical may be administered by mouth, applied to the skin, or both. The subject animals can be albino rabbits or rodents, and the challenge test is exposure to ultraviolet A (UVA) and ultraviolet B (UVB) light, or mechanically generated light covering the same wavelengths.

Chronic Toxicity

The purpose of conducting chronic toxicity experiments is to obtain information about the possible or probable adverse health effects that could result from long-term exposure to relatively small quantities of chemicals. There are two levels of concern about the chronic toxicity of chemicals. One concern is for health effects in a population of young and old, healthy and infirm, men and women, exposed to a wide variety of chemicals in the environment or to medicines that are taken for a long time, such as blood pressure or diabetes medicines that may be used for 50 years or more. The other is for health effects in people exposed to chemicals in their occupations. The latter exposures are fewer in number for any one occupation, but the quantities of exposure, while still small, are up to tens of thousands times greater than environmental exposures. The study of the chronic effects of chemicals is a complicated, lengthy, and expensive process.

How is the chronic toxicity of a chemical studied? Experimental protocols for chronic toxicity testing are many and varied, the methodology depending on such factors as route of exposure (oral, intravenous, dermal, inhalation), nature of effect (organ damage, mutagenicity, carcinogenicity, birth defects, etc.), and the populations of concern (adult, child, developing fetus, etc.) which are important determinants of the experimental design.

A wide variety of animal species are used in the study of the chronic toxicity of chemicals, ranging from monkeys to dogs, cats, pigs, and numerous rodent species, down through nonmammalian animals, such as birds, reptiles, amphibians, and fish. The most common test animals for chronic toxicity studies are rats and dogs. Nonhuman primates may be used extensively to test certain types of medicines, especially biologics where human proteins would lead to immune reactions in other species, but where these reactions are limited in monkeys due to their closer relationship to humans. Some types of toxicologic experiments use microorganisms, tissue cultures, or isolated animal organs. These provide valuable information and direction for further investigation, but with our current state of knowledge, they cannot yet replace whole animals in the study of chronic toxic effects of chemicals.

The classic chronic oral toxicity experiment is quite simply an animal dosing study in which the animals are administered the chemical daily. Before such a study is begun, range-finding experiments based on a knowledge of acutely toxic doses of the chemical under investigation are undertaken to determine what daily doses the animals would be able to tolerate for prolonged periods. These range-finding (subacute) studies may be from 5 up to 90 days in duration, depending on how long and in what population the chronic studies will be conducted. Based on the data obtained from these investigations, three or more dosages of the toxicant are selected for the chronic study.

For chemicals found in the environment, the toxicant may be mixed with animal feed to give the desired concentration for each dosing level. The highest level fed to the animals is the one calculated to have significant chronic toxic effects. If the high-dose level kills the animals before the end of the experiment, its value in providing meaningful chronic toxicity data is limited. Therefore, the choice of the high dose is critical to the usefulness of the study. The lowest level fed to the animals is, ideally, one that will produce no detectable adverse effects. The intermediate level(s) are those that will produce effects intermediate between the high and low levels. One group serves as a control, an important part of most toxicity study protocols. Control animals are treated in exactly the same manner as the experimental animals, except that the test chemical is not added to their feed. Control animals are essential in long-term studies to ensure that any adverse effects that occur during the course of the experiment are, in fact, due to the test article and not to some other condition of the experiment.

For other types of chemicals, including drugs, the doses are administered by the route expected to be experienced by humans (oral, intravenous, dermal, or inhalation). In these studies, the doses are calculated to give the same level of toxicity as in feeding studies and, again, dose-ranging studies of shorter duration must be performed in order to determine the correct doses for the chronic studies. Currently, FDA requires a 6-month rodent study and a 6-, 9-, or 12-month nonrodent study for the registration of drugs that will be taken for greater than 3 months by humans. Costs for each of these studies can reach upward of $1 million and generate data that may take 2,000 pages to summarize. Feeding studies that may last up to 2 years may cost even more. These costs exclude the cost of the compound being administered, which may be relatively small for some classes of compounds but extremely high for medicinal compounds such as synthetic human proteins.

Relatively large numbers of animals are assigned to each exposure group in order to anticipate some mortality either due to the aging of animals, rodents in particular, or effects of the compound. Some attrition can be expected, even in control groups. When rats are the experimental animals, perhaps 20 females and 20 males will be assigned to each group. When dogs, swine, or monkeys are used, the groups are comprised of a minimum of 4 females and 4 males each. In addition, recovery groups are sometimes included to assess the ability of the organism to resolve any toxicity after cessation of exposure. Animals are observed daily for any signs of abnormal health or behavior. All animals are given a more detailed examination weekly, at which time they are weighed, their feed consumption and sometimes water consumption is measured, and any scheduled clinical tests are performed. During the experiment, various other biochemical and clinical tests may be performed periodically in an attempt to determine if there are any adverse effects that are sufficiently subtle to escape gross observation. At the end of the experiment, all animals are necropsied, where they are examined grossly for any signs of lesions including tumors or other pathologic change. Selected organs are weighed for evidence of atrophy or hypertrophy, and numerous tissues are preserved for microscopic examination for evidence of histopathology. All data are evaluated statistically to determine if there is a relationship between dose and effect.

Any discussion by science or industry of the great costs of chronic toxicity testing should not be viewed as a tacit argument against such investigations. Rather, it is a statement of why such testing is, for the most part, limited to chemicals that promise to have sufficient commercial value to justify the costs. The high cost of toxicologic evaluation is a fact about which the public is generally unaware but which exerts considerable influence on decisions that affect their health and well-being.

No matter how much time, effort, and money are put into the study of the chronic adverse effects of chemicals, and no matter how many studies yield negative results, one can never be sure that there is not some subtle effect yet to be discovered. Each chronic toxicity experiment adds to the fund of knowledge about long-term effects of chemicals. As we develop a larger and larger data base of chronic effects of chemicals, nontoxic as well as toxic, and as our knowledge of mechanisms of action of chemicals increases, the need for classic chronic animal studies may diminish. Before that time arrives, other problems, such as the chronic toxicity of chemical combinations and the significance of synergistic and antagonistic chemical interactions, will have to be dealt with.

Mutagenesis

The purpose of mutagenicity testing is to study the ability of chemicals to alter the genetic code—that is, to determine if they are mutagens. A number of different methods are used to investigate mutagenesis. Some involve microscopic examination of the nuclei of cells, such as white blood cells or the cells of other tissues, to see if the chromosomes are abnormal in size, shape, or number. Some mutations involve only a subunit of a chromosome and do not alter the gross appearance of the chromosome on which it resides. At the present time, there is no way for geneticists to determine if genes are mutated by looking at them directly. Such mutations can be detected only indirectly when they express themselves in the mutated organism or its progeny.

Mutagenicity studies employ a variety of test organisms, such as rats, mice, or other mammals, animal tissues, insects, or microorganisms. No one test method can adequately describe mutagenic risk for humans. Thus, it is usual for batteries of tests to be carried out when evaluating potential mutagenicity of chemicals for humans.

One test that is commonly used for preliminary screening of chemicals for mutagenicity, the Ames test, is named for Bruce Ames, the scientist who developed the technique in the 1970s. The organism used in the Ames test is *Salmonella typhimurium*, a microorganism that has lost its ability to synthesize the amino acid histidine as a result of mutation from its normal wild type. This mutant requires the presence of histidine in its media for it to survive and multiply. In the Ames test, the mutant organism and the test chemical, plus biochemical activators, are placed in media lacking histidine. A certain number of the mutants will spontaneously revert (mutate) back to the normal wild state and will be able to grow in the absence of histidine. If the test chemical is capable of causing mutations in the organism, the number of organisms that revert back to the wild type will be greater than the number that spontaneously reverts. The greater the number of reversions, the greater is the potency of the mutagen.

It is usual to study compounds in the Ames test (also referred to as a reverse bacterial mutation assay) to look at changes to DNA, in some type of in vitro test to look at changes in chromosomes, and in an in vivo test in mouse or rat in order to see if in vitro changes are predictive of the in vivo condition.

Any chemical that is judged to be a potential human mutagen is one for which human exposure should be carefully monitored or controlled. However, it is ironic that the test most commonly used to screen for mutagenicity is a test for a chemical's ability to convert a mutated organism back to its normal state.

Not all mutations involve the DNA of an organism. Epigenetics is the study of changes that occur outside of the DNA. Certain environmental factors seem to alter the expression of characteristics so that they are able to be inherited by offspring much in the same way that genetic changes to DNA are inherited. For example, fruit flies (the much-beloved experimental animals of geneticists) that have white eyes when raised at 25°C pass along white eyes to their offspring. When the temperature is raised to 37°C for a brief time, they have offspring with red eyes, which characteristic is passed along to their offspring. The area of epigenetics is undergoing rapid research and may end up explaining some human genetic traits.

Carcinogenesis

Because mutation is considered to be a critical event in cancer causation, screening tests for carcinogenic potential of chemicals employ some of the same methods as those used in mutagenicity testing. These screening tests give valuable information about carcinogenic potential. However, the ultimate test must be exposure of test animals to the chemicals in question for an extended period of time. The reason for this is that some chemicals that do not yield positive results in mutagenicity tests do cause cancer in animals while others that are mutagens are not carcinogens.

One of the shortcomings of animal carcinogenicity testing is related to the long induction periods and low incidences associated with exposure to small quantities of carcinogens. If there is any exposure of the general public to suspected carcinogens in air, water, or soil, it is usually very small, measured in parts per billion (ppb) or parts per trillion (ppt). The use of such small exposure levels in laboratory experiments would require astronomical numbers of animals in order to detect a carcinogenic effect. Since such studies are not feasible, the common practice in animal carcinogenicity testing is to administer large doses, tremendously larger than would be encountered by the general public, in order to increase the potential for demonstrating a carcinogenic effect of the test chemical. Maximum tolerated doses, which may be very large, are the highest nonlethal doses the animals can tolerate for the duration of the experiment.

The use of these high doses is accepted by regulatory agencies despite acknowledged pitfalls of which the public is generally unaware. These pitfalls derive from the fact that the biochemical fate of very small doses of a chemical may not be the same as that for large doses. The difference in effect that chemicals display between their acute and chronic toxicities is testimony to that fact. Small doses of a chemical may follow a metabolic pathway that does not convert it to a carcinogen. With increasing doses the pathway becomes saturated, and the excess chemical is diverted to a different pathway that converts it to a carcinogen. Or small doses of the chemical may be prevented from exerting carcinogenic activity when combined with a biochemical normally present in the body. If the supply of the biochemical is exhausted by large doses, the excess chemical is then free to exert its carcinogenic effect. The only chemicals for which effects at high levels would accurately represent low-level effects are those whose biochemical fate is not altered by dose.

Some analogies to the use of high doses, ridiculous though they may seem, may give the reader an idea of the kind of distortions produced by animal tests at high doses. A team of sports physicians at a hypothetical university is interested in investigating the adverse effect, if any, on the ankle, knee, and hip joints of athletes who participate in pole-vaulting events. The problem is that it would take hundreds of thousands of pole vaulters making their jumps every day for a period of many years to obtain sufficient data that could produce statistically significant results. Since such an experiment is impossible, it is decided that instead of 100,000 people making a 20-foot jump each day for many years, 1,000 people will make a 200-foot jump 10 times a day for one year. Since no athlete could vault 200 feet, and since the trip up is of no importance to the experiment, a nearby 200-foot cliff is selected as the jumping-off place.

This analogy grossly exaggerates an indifference to the obvious biologic limitations of the test organisms by suggesting conditions that are greater than the maximum that could be tolerated. Nevertheless, carcinogenicity testing using heroic doses may result in similar indifference.

Proponents of the use of maximum tolerated doses justify their position by saying they do not artificially induce cancer. They contend that if a chemical causes cancer in very high doses, it will also cause cancer in very low doses. This position denies the fact that changes in metabolic pathways of a chemical can occur with increases in dose. Knowledge of biochemical mechanisms and data provided by study of the metabolism of carcinogens belies the accuracy of the maximum tolerated dose. The use of maximum tolerated doses in animal carcinogen studies does provide valuable information and certainly should not be deleted from testing protocols. However, acceptance of results from high-dose exposures as the only valid data, with concomitant rejection of data from moderate-dose studies and studies of mechanisms of action, metabolic fate, and so forth represents an attitude that is foreign to objective scientific inquiry and does a disservice to the public.

A more moderate approach to carcinogenicity testing is the use of doses comparable to those received during exposure to chemicals or a modest increase over this exposure. Working populations, as a rule, do not include young children, senior citizens, or people who are ill or debilitated, but their lack of representation in occupational groups is compensated for by the fact that occupational exposures to chemicals are usually several to many thousands of times greater than those encountered by the general public. Occupational exposures are sufficiently high to provide valid data from a number of animals that can be reasonably accommodated in a toxicologic laboratory. This approach, which is employed by some research scientists investigating carcinogenicity, uses exposure levels that reflect those found in occupational settings; a chemical that does not cause cancer in occupationally exposed people is very unlikely to cause cancer in the general public, particularly at dose levels that are thousands of times lower than those found in occupational settings.

Carcinogenicity testing for medicinal compounds is usually performed with doses selected jointly by FDA and the commercial sponsor. The high dose should be one where the animals show some slight toxicity in range-finding studies but where that dose would not lead to early deaths in the study. Since standard carcinogenicity studies usually are considered lifetime studies in that they last for up to two years in mice and rats, some animals will die of old age. Therefore, the number of animals in each group is very high, and so of course is the cost. Because of the cost involved, carcinogenicity studies are usually not done until medicines are shown to be effective in a small group of people.

In addition to the long-term carcinogenicity bioassays, there are several six-month studies available in genetically altered mice that can be substituted for the two year studies. These studies involve the use of mice for which tumor suppressor genes have been "knocked out" or tumor promoter genes have been "knocked in," that is, where the animals have been made more susceptible to the effects of carcinogens.

Developmental and Reproductive Toxicity

The original methods for the study of developmental and reproductive effects were designed by nutritional scientists in the early 1930s. Method development was stimulated by numerous observations that maternal malnutrition could exert a profound effect on the fetus and the young in the postnatal period. Nutritional studies were gradually extended to investigations of reproductive effects of excess nutrients or other chemicals added to the maternal diet.

Three basic types of studies were developed to study effects of chemicals on reproduction and development. These studies are referred to as the DART (developmental and reproductive toxicology) studies. In the first type (referred to as a Segment I study), the test chemical is administered to both males and females prior to mating and to females continuing until the young are weaned. The second involves administering the test chemical to a female after pregnancy has been established. This study is sometimes referred to as a Segment II or embryo-fetal study. This type of study is a tool for teratologists to study fetal malformations (i.e., birth defects). In the third type of study, the test chemical is administered to the pregnant mother through weaning, and after-weaning exposure to the mother and some of the offspring is continued until two, three, or more generations are produced.

Multigeneration tests were conducted on any new chemical that might find its way into the food supply, either deliberately or inadvertently. In multigeneration studies, groups are set up and examinations are conducted in the same manner as described for chronic toxicity testing. When the animals reach maturity, they are bred and permitted to deliver and wean their young before they are submitted to the necropsy examination. Pups from each litter are selected to become the parents of the next generation, and the remaining pups are necropsied. This process is repeated until two, three, or four generations are produced. This type of experiment provides information about whether chronic exposure to the chemical in question adversely affects the overall reproductive process or produces some condition that does not express itself prior to the second or third generation.

For many drugs that will be given to people of reproductive age or earlier, all three segments of the DART studies usually are required before the drug is approved in the United States. Package inserts for approved drugs contain a pregnancy category, A, B, C, D, or X. Category A means that studies in pregnant women show no fetal risk. Since this category requires that pregnant women be included in clinical trials, very few drugs fall into this category. A category B drug is for medicines for which there are no studies in pregnant women but the DART studies show no fetal risk, or in some rare cases where animal studies showed some risk but human studies did not. Category C drugs are those for which there have been no human studies, animal studies show fetal risk, but the risk-benefit ratio is acceptable. A category D drug is one where there is human fetal risk based on postmarketing surveillance but the benefit outweighs the risk (i.e., it is important for the pregnant woman to control her disease regardless of the risk to the fetus). Category X drugs have shown both animal and human fetal risk, and the risk outweighs the benefit. Women who are pregnant or who expect to become pregnant should discuss with their physician before taking any medicine the potential risks of *any* drugs, prescribed or over the counter,

In recent years, many more types of tests have been developed to study the nature and location of effects of chemicals in male and female reproductive processes. Some of these tests involve the use of live animals; others involve the use of animal tissues or cells. The techniques employed come from a variety of disciplines, including physiology, biochemistry, genetics, and molecular biology.

UNITS OF TRACE QUANTITIES

The terms *ppm*, *ppb*, and *ppt* are in common usage. People know that *ppm* means "parts per million," *ppb* means "parts per billion," and *ppt* means "parts per trillion." But what do they really mean? How much is one unit of some substance in a trillion units of some other substance (ppt)? What is its significance? These questions take on even greater importance when one recognizes that analytical chemists are pushing the limits of detection farther and farther down in the scale of quantification. We must be prepared for ppq (parts per quadrillion), ppp (parts per pentillion), and so forth.

How much is one part per million? One part per million is equal to 1 drop in 14 gallons. A ppb is 1/1,000th of a ppm. Therefore, a ppb is equivalent to 1 drop in 14,000 gallons. A ppt is 1/1,000th of a ppb. Therefore, a ppt is equal to 1 drop in 14,000,000 gallons. Or, as a colleague once remarked,

"A very dry martini!" Other colleagues have offered other analogies. A ppm may be likened to the diameter of one hair being expanded to the diameter of the Holland Tunnel. A ppt is equal to the thickness of a dollar bill in a stack of bills 63,000 miles high. All sorts of similar calculations can be made to dramatize the extremely small quantities that ppm, ppb, and ppt represent.

Even though ppm, ppb, and ppt are extremely minuscule quantities, they still represent billions and billions of molecules, as can be seen from the discussion of benzo[a]pyrene in Chapter 7. Recognition of this concept is necessary for an understanding of the difficulties inherent in requiring that the quantity of any chemical contaminant in any medium—air, water, food, soil—be zero. Zero, the absence of even one molecule, cannot be measured. Therefore, zero contamination has no practical meaning for any chemical that has any commercial use at all.

The only chemicals that we can say with reasonable certainty are present in the environment in zero amounts are chemicals that have never existed. For all others, natural and synthetic, the best that analytical laboratories can do is determine that they are not present in amounts detectable by the most sensitive analytical equipment available. The quantity of a chemical that is below the level of detectability depends on a number of factors, such as the nature of the chemical in question and the medium in which it occurs. The level of detectability becomes smaller and smaller with improvements in analytical techniques, but it will probably never reach zero.

ANALYTICAL METHODS

The public, in general, has complete faith that the concentrations of contaminants reported in environmental samples, wildlife species, food, drinking water, drugs, and most especially in organic or pesticide-free produce are accurate and absolute. People who are not chemists do not appreciate the tremendous technical difficulties of measuring ppb or ppt of any chemical in any medium. Analysis of such extremely small quantities requires elaborate and expensive analytical equipment that is not to be found in just any laboratory. Further, any laboratory that is capable of accurately measuring ppb or ppt quantities must be impeccably clean; even ppb contamination of solutions, reagents, laboratory air, or the most minuscule traces on glassware, bench surfaces, or equipment would completely obscure the presence of ppb or ppt in a sample. A contaminated laboratory can only give unreliable results. More often than not, a contaminated laboratory will report the presence of a trace amount of chemical when it is actually not present at all in the sample. Concentrations of chemicals in the ppb and, more especially, the ppt ranges are invisible in many laboratories, too small to be detected by analytic instrumentation and methods.

Some laboratories will report an unqualified zero when an analysis does not detect any of the chemical in question. This is a meaningless report because no laboratory can determine zero (that no molecules of the chemical are present). Some laboratories, however, will give a figure such as "<0.2 ppm," which gives the impression that some of the chemical is present, perhaps as much as 0.199 ppm, when in actual fact all it really says is that the chemical was not detected, but that if it was present it was in a concentration of less than 0.2 ppm. The majority of laboratories report results in the proper manner, which is to state the quantity found (or zero, if none was detected) followed by the sensitivity of the analytical method— the smallest concentration that is capable of being detected by the analytical procedure, referred to as the LLOQ, or lower limit of quantitation.

Another problem with analytical procedures for some of the very complex organic chemicals is whether they really detect what they are designed to detect. Whole groups of chemicals can give the same response in some procedures. Without methods for separating the various components of the group, there is no way of knowing which or what combination of them is present in a sample. This was the problem with the early analyses for DDT in environmental samples. Even today, chemical analysis for trace quantities of chemicals is an extremely difficult and tedious procedure, although with the advent of mass spectrometry, the true identity of the analyte (the compound being tested) can usually be ascertained.

The question of the significance of the results of chemical analyses should be of particular concern to people who want and demand pesticide-free foods. Consumers should ask what chemicals are detected by the methods used, how sensitive the methods are, and the competence of the laboratory employed to test their produce. The health effects of exposure to trace quantities of environmental chemicals are independent of whether those quantities are detected or undetected.

People also seem to have the impression that analytical laboratories have much greater capabilities than they do. This has been reinforced by various television shows that glamorize the crime scene technicians and show them, week after week, correctly identifying all kinds of trace materials. Real life is not that easy, nor are there monetary resources to identify every compound. As an example, many people who feel they have been made ill by some food or other material often want to send it to a laboratory to determine the identity of the offending contaminant. They do not realize that laboratories can only analyze for specific chemicals or specific groups of chemicals.

Laboratories cannot just look for anything: The chemical for which the analysis is to be made must be specified. In light of the fact that there are hundreds, if not thousands, of potential contaminants, an analysis of a sample for an unspecified chemical would be a major and extremely costly research project.

Some final thoughts about laboratory analyses: If a physician suspects acute or chronic poisoning from some chemical, such as lead or arsenic, clinical laboratories can provide very valuable and necessary information to assist in the diagnosis. In such instances, the concentration of the suspect chemical in body fluids or tissues is very much greater than the concentration that would occur from trace environmental exposure. Deliberate poisoning is one of the easier tasks for the forensic toxicologist; looking for the effects of long-term chronic low-level exposure to an environmental contaminant is considerably harder.

More commonly, a person concerned about environmental chemicals in the body wants to know where to obtain an analysis for the presence of some synthetic chemical. Often the people most concerned are young mothers who want an analysis for some pesticide or other persistent chemical (e.g., polychlorinated biphenyls [PCBs] or dioxins) in their breast milk. We now know that people are exposed knowingly or unknowingly to a variety of environmental chemicals, many of which will find their way into breast milk of both humans and animals. Before testing, the person should ask, "What am I going to do with the information, once I have it?" The usual answer to this question is that they do not know. Often they are not aware that, in the absence of any unusual exposure to the chemical, a fairly accurate guess can be made as to what the analytic results would be produced without performing a laboratory analysis. It is important that mothers know that there are no known adverse effects in nursing babies from mothers' milk containing trace quantities of most environmental chemicals. In addition, it is generally accepted that the immunologic and psychological benefits of nursing usually far outweigh any unknown, subtle effect hypothesized for such chemicals.

An interesting point that may help give a perspective to today's young mothers is that if they themselves were breast-fed babies, the chances are very good that the milk they received from their mothers had higher concentrations of such chemicals as DDT, PCBs, and dioxins than their own milk does now. Those chemicals are so much more rigorously controlled today than they were several decades ago that human exposures today are much lower. However, other chemicals appear to be increasing in human milk today, such as the flame retardant polybrominated diphenyl ethers (PBDEs). Studies are under way to determine the long-term effects of exposure to these compounds.

Every one of us has trace quantities of chlorinated hydrocarbon chemicals stored within us—DDT, DDE, PCBs, dioxin, and so on. No known clinical significance can be attached to trace quantities of these chemicals, or even to higher-than-average quantities of storage. Thus, there is no practical value in spending money for laboratory analyses to determine one's own body concentrations of these chemicals.

ANIMAL RIGHTS

No discussion of toxicity testing methods would be complete without reference to the animal rights movement. The great increase in the use of laboratory animals that accompanied the commercial development of synthetic organic chemicals after the end of World War II was followed by increased public concern for the humane treatment of laboratory animals. Apart from their concern for the welfare for the animals in their charge, scientists who used animals in their research long ago recognized that maltreated or sickly animals were poor subjects because they gave unreliable results. In order to avoid such difficulties, scientists and their member societies gradually developed protocols for laboratory animal care. These protocols eventually became incorporated into federal law in 1967 as the Laboratory Animal Welfare Act.

Laws and regulations requiring humane treatment of laboratory animals never could be sufficient for people totally opposed to the use of animals in research. The fact that laboratory animals are unwitting participants in research, combined with the assumption that such research is always cruel and painful, gave rise to the animal rights movement. This movement, as with any involving social issues, represents philosophies that range from moderate to extreme. Moderate views recognize that advances in medical and allied research are dependent on the use of animals for some procedures. They advocate the development of methods that do not require living creatures and, when no alternative research methods are available, that experimental animals be treated in the most humane way possible. These goals are also the goals of science and of most of the scientists who use animals in research.

Groups with extreme views have targeted animal experimentation for extinction. These views have led to extremist tactics, such as destruction of laboratories and research records, liberation of laboratory animals, harassment of researchers, and threats and attempts against the lives of research scientists and industry personnel. Unfortunately, those who suffer most from these criminal acts are the very animals that the acts are intended to help.

Whenever research records are destroyed, the animals who participated did so in vain. Such research must be repeated, requiring the use of more animals.

The liberation of experimental animals is the cruelest act of all. Animals bred for research are not feral animals capable of surviving against all odds in the wild. They do not have the adaptability of their wild counterparts, and they suffer greatly during their short period of survival in the wild. Because they have been bred in captivity for many generations, they have become totally dependent on their keepers and have lost survival instincts. They do not know how to find food or safe shelter. They do not know how to protect themselves from harm, and they have no resistance to the diseases common to their wild fellow creatures. Some years ago animal rights extremists "liberated" an entire mink farm. Although these animals were not used for research, this release demonstrates how hard it is for semidomesticated animals (which might have an easier time in the wild than animals bred for laboratory use) to live in the wild. Of the approximately 5,000 mink released, fewer than 300 animals could be found during the next few weeks, while thousands of carcasses were found over the next year, either dead from starvation or killed by dogs, coyotes, and other predators, or hit by cars. The mink farm went out of business, which was probably the real mission of the "liberators," but most of the animals suffered in the process—not a good outcome for people who claim that their objective was concern for the welfare of the animals.

Today, animal-based research is essential to protect the health and well-being of humans. Animal experimentation is conducted, almost without exception, for the prevention or alleviation of human ills and suffering. Toxicologists would prefer to have nonanimal methods available for their research because animal subjects are extremely expensive to buy, maintain, and use. Despite the claims of some animal rights activists, scientists who use animals for research are not sadists who enjoy inflicting pain on their subjects. On the contrary, they have concern for the quality of life of their laboratory animals and the prevention of pain and suffering in them. There is considerable interest in finding alternative nonanimal methods of study, and numerous investigations of nonanimal methods are in progress. However, it is very difficult to find a nonanimal substitute for a human tissue or organ, much less the whole organism. If humans were smart enough to program computers to display the complexity of the human body's interaction with chemicals, researchers would jump at the chance to omit animals from their research paradigm. After all, computers don't bite, they don't have to be fed, they don't need their cages washed, and the researcher could go on vacation without worrying about the status of his or her experiment. Unfortunately, with our current knowledge of physiology, pharmacology, and toxicology, we are unable to program in silico experiments using computer models instead of animals that can mimic the in vivo experience.

Perhaps more important is the problem of validation. Once a nonanimal method is found that gives promise of being suitable, how do results obtained with it compare to those from animal studies? How do data from nonanimal systems translate to effects in humans? The answers to these questions must be obtained before a new nonanimal method can be substituted for an older, proven animal method. The process of validation requires a great deal of time, effort, and money. In addition, comparison between the nonanimal and animal method requires many animals in order to ensure that the outcome of both methods is similar as well as predictive.

Despite great activity in the search for alternate methods, there are, at the present time, many kinds of studies for which no alternate methods are available. In fact, it is conceivable that a substitute for the whole animal may never be found because most research is dependent on the interplay of the tremendous number of complex physiological and biochemical processes and interactions in the living organism.

CHAPTER 6

GENERAL TOXICOLOGY

The previous chapters have defined the various types of toxicology studies that are performed. Chapters 6 through 8 discuss the usefulness of these studies as well as the types of information that can be derived from each.

ACUTE TOXICITY

Acute toxicity refers to the ability of a substance to do systemic damage as a result of a one-time exposure of relatively short duration, so called single-dose toxicity. Such exposures are frequently accidental in nature. A great deal is known about the acute oral toxicities of a great many chemicals. Experiments designed to determine acute oral toxicity are relatively easy to perform and are not very expensive.

LD_{50} and LC_{50} Values

As discussed previously (Chapter 5) LD_{50} is the technical term used to describe the acute oral, intravenous, or dermal toxicity of chemicals. LC_{50} is used to describe acute inhalation toxicity. LC_{50} is also used to express the toxicity of chemicals to fish or other aquatic organisms. In aquatic studies, LC refers to "lethal concentration" of a chemical dissolved in water. The subscript refers to the percentage of the animals for which the dose was lethal. For example, a subscript of 50 means 50 percent of the animals died, 100 means 100 percent of the animals died, and 0 means no animals died. LD_{50} and LC_{50} values are read from plots such as that shown in Figure 6-1. The smaller the LD_{50}, the greater the toxicity of the chemical, and conversely, the larger the LD_{50}, the lower the toxicity. This inverse relationship between LD_{50} and degree of acute toxicity can be confusing unless or until one becomes familiar with the use of LD_{50}.

The Dose Makes the Poison: A Plain Language Guide to Toxicology, Third Edition.
By Patricia Frank and M. Alice Ottoboni
© 2011 Patricia Frank and M. Alice Ottoboni. Published 2011 by John Wiley & Sons, Inc.

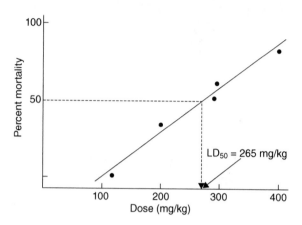

FIGURE 6-1   Acute dose-mortality (LD_{50}) curve.

For a variety of reasons, there are many more oral, intravenous, and dermal LD_{50} values recorded in the scientific literature than there are LC_{50} values. Many chemicals do not become airborne easily. For these chemicals, often it is difficult or impossible to achieve air concentrations that are harmful, much less lethal. For chemicals that do become airborne easily, the technical difficulties of administering known and constant concentrations of gases, vapors, or dusts to groups of animals in environments isolated from laboratory personnel have been overcome, but the process is more expensive than for other routes of administration.

Today, in an effort to use as few animals as possible in toxicology experiments, a full LD_{50} determination may not be performed. However, the concept of the LD or LC value is helpful for understanding acute toxicity. Methods for determining LD values as well as other assessments of acute toxicity are discussed in Chapter 5.

Significance of LDs for Humans

An important source of information about the acute toxicity of chemicals for humans comes not from LD_{50} values but, unfortunately, from deliberate and accidental poisonings. When homicides, suicides, or accidents occur, the estimated quantity of chemical to which the victim was exposed usually can be made. That quantity can then be related to the outcome—that is, severity and kind of symptoms, survival, or death. For many chemicals, human experience corroborates animal toxicity data. If a chemical is shown by data

TABLE 6-1:   Relationship between Animal LD_{50} Values and the Quantity of Chemical for Human Lethality

LD_{50} (mg/kg)	For a 10-kg Child	For a 70-kg Adult
Up to 5	Up to 1 drop	Up to 1/16 teaspoon
5 to 50	1 drop to 1/8 teaspoon	1/16 to 3/4 teaspoon
50 to 500	1/8 to 1 teaspoon	3/4 to 3 tablespoons
500 to 5,000	1 teaspoon to 4 tablespoons	3 to 30 tablespoons
Over 5,000	Over 4 tablespoons	Over 30 tablespoons

from humans to be more or less toxic to humans than to laboratory animals, human data must take precedence.

LD_{50} values sometimes are used to estimate lethal doses for humans; however, when doing so, it must be recognized that they are based on animal experimentation. Reason and judgment must be used to translate their meaning to human terms. When estimating the quantity of chemical that would be potentially lethal for a child or adult using an LD_{50} value, multiply the weight of the person (in kg) by the LD_{50} value (in mg). A 22-pound child weighs 10 kg, so the LD_{50} is multiplied by 10, yielding a dose that would be toxic to half the children. For example, potentially lethal doses of a chemical with an LD_{50} of 10 mg/kg would be 100 mg for a child and 700 mg for a 70-kg adult. The relationship between oral LD_{50} values for animals and the quantities of chemical they would represent for children and adults is shown in Table 6-1.

Table 6-1 clearly demonstrates that children should be protected from acute exposure to nontherapeutic chemicals, regardless of how innocuous they may seem to be for adults, and to the lowest dose possible for therapeutic chemicals. A chemical having a probable lethal dose of about 1 pound (approximately 30 tablespoons) for an adult would require only about 4 tablespoons to be lethal to a child—not a very large quantity. Excluding differences in sensitivity to chemicals because of age, lethal doses of chemicals are smaller for children than for adults because children are smaller in body weight. Differences in body weight also account for the fact that insects, with body weights that are minuscule compared to those of humans, are killed by very tiny amounts of insecticides, amounts far below those that are acutely harmful to humans. Compounding lower body weights in children may also be differences in the metabolism of chemicals by the pediatric population. In recognition of this, the Food and Drug Administration (FDA) routinely requests pharmaceutical companies to conduct studies in juvenile animals in order to determine the toxicity of these chemicals in an applicable model for children. However, remember that this is a relatively new concern of FDA, and there are many drugs on the market for which no pediatric studies have been conducted. Doses of these drugs usually are estimated by scaling down the safe and effective dose for adults to the lower body weights of children, not taking into account the differences in the patient populations.

It should be mentioned here, if it has not already been made obvious, that the LD_{50} or LC_{50} for humans is not known for any chemical. LD and LC values are determined by very specific protocols under controlled laboratory conditions. For obvious reasons, such experiments could not be conducted with humans. Nevertheless, human LD_{50} and LC_{50} values are given for a number of chemicals in some toxicology reference books. What these publications actually refer to are not true human LD_{50} values but rather average lethal doses (ALDs) or mean lethal doses (MLDs) calculated from accidental or homicidal deaths.

How are LD_{50} data from animals used to estimate the acute toxicity of chemicals for humans? Where there is great concern about human toxicity from a new chemical because it has a potential for extensive human exposure, routine LD_{50} studies in rats and mice are supplemented with studies performed on additional species, such as guinea pigs, dogs, and monkeys, in an attempt to determine species variability in response to the chemical. Some chemicals are quantitatively similar in acute toxicity among species and others vary widely. As a general rule, if a chemical has the same degree of acute toxicity for all species tested, it probably will have a similar toxicity for humans. If data from different species of animals vary widely, an estimate of where people may fit in the scheme is difficult to make, and further information about how the chemical exerts its toxic action is required. However, regardless of best estimates, it is always assumed that humans are more susceptible to the adverse effects of chemicals than the most sensitive species tested until or unless there is reliable evidence to the contrary.

The slope of the dose-mortality curve provides information that is useful in evaluating potential acute toxicity of chemicals for humans. It may be steep (see Figure 6-2(a)), or it may be shallow (see Figure 6-2(b)). What is the significance of the slope? It provides information about how variable individuals within a species are in their responses to the chemical in question. It tells when the lethal dose range is narrow (steep) or wide (shallow). When it is steep, within the species there is only a small difference between the dose that is lethal for the most susceptible animal and the dose that is lethal for the most resistant animal. When it is shallow, there is a wide difference among the lethal doses. Even though a chemical may show large variations

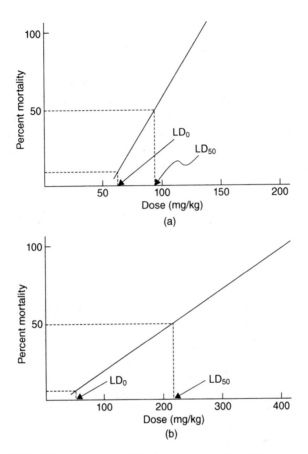

FIGURE 6-2   (a) Steep LD_{50} curve; (b) shallow LD_{50} curve.

in LD_{50} values among species, the dose-mortality curves are usually quite similar: If it is steep for rats, it probably will also be steep for mice, dogs, monkeys, and so on unless there is species-specific metabolism.

If the dose-mortality curve is steep, humans probably will also show little individual variation in the quantity that will produce an adverse effect. The organophosphate pesticides are chemicals that as a class have steep dose-mortality curves for laboratory animals. Experience shows that humans also have little variation in their responses to this group of chemicals. For many organophosphates, doses of up to a quarter or a third of the LD_{50} dose will produce no deaths at all (see Figure 6-2(a)). A steep dose-mortality slope gives much more assurance than a shallow slope in estimating safe acute doses for humans. Where human life is involved, the concern is for safe doses, not doses that will kill 50 percent of the people exposed!

If the dose-mortality curve is shallow, a small fraction of the LD_{50} value will be lethal to some individuals (see Figure 6-2(b)). If test animals show such variation, humans probably would also. In such cases, we cannot predict how an individual human will respond to the chemical. Therefore, we must assume that all humans are as sensitive as the most sensitive person. In this age of genetic testing and the ability to determine DNA sequences, it is becoming possible to predict whether a person will respond to a given medicine. This same technology can predict susceptibility to toxicity as well; however, it would not be economically feasible to screen individuals on the chance they might be exposed to an environmental chemical.

An example of a chemical with a shallow slope is diethylene glycol, one of the chemicals used to formulate antifreeze. As little as a tenth of the LD_{50} dose is lethal to some animals. Extrapolation of such data to humans is difficult. The episode involving elixir of sulfanilamide described next tragically demonstrated that dose-mortality curve of diethylene glycol is also shallow for humans.

In 1937, the wonder drug sulfanilamide had just become available for general medical use in the treatment of bacterial infections that had almost always been fatal. One formulation was dispensed as an elixir of 10 percent sulfanilamide, 72 percent diethylene glycol, and the balance flavoring and coloring substances. Diethylene glycol was selected because it was a good solvent for sulfanilamide. There was little information in the scientific literature at the time relating to the acute toxicity of diethylene glycol, and the pharmaceutical company that formulated the drug was unaware of its potential danger. Shortly after the elixir was put on the market, reports began appearing in newspapers around the country about deaths from the new wonder drug. The American Medical Association was deluged with calls from physicians wanting information on the causative agent. Was it the sulfanilamide, the vehicle, or some contaminant? The use of sulfanilamide was totally suspended. Within a short time, based on animal studies, diethylene glycol was found to be the offending agent. Diethylene glycol was immediately banned as a vehicle for drugs, but not before 100 or more people became ill and at least 76 people had died from the elixir. Some of those who survived had taken as much as 10 oz of elixir, whereas some of those who died had taken as little as 1 oz.

It is sad to report that another outbreak of poisoning occurred in the twenty-first century when acetaminophen for children was mixed with diethylene glycol in four batches of teething medication. This material was sold in Nigeria over a six-month period and resulted in 54 deaths from renal failure, which was over 95 percent of the children exposed (J. Schier, National Center for Environmental Health, CDC).

Poison Prevention

Fortunately, mass accidental poisonings, such as occurred in the diethylene glycol episode, were uncommon in the past. They are even more unlikely to occur now in the United States because of the greatly tightened controls built into food and drug regulations. Recent episodes, such as the deliberate addition of melamine to milk and pet food and the resulting deaths to dogs and cats in this country, as well as deaths and kidney failure to many children in China, raise the issue of the safety of our food chain. Public outcry to make our food safe from poisons, pesticides, bacteria, and other contaminants probably will lead to stricter control in the United States and Europe, but people residing in developing nations will still be at risk.

Today, young children are the victims in the great majority of cases of acute poisoning from accidental ingestion of chemicals. The World Health Organization estimates that 45,000 deaths worldwide occur annually in children with most occurring in children under one year old. The tragedy in these statistics is that most, if not all, of these unfortunate events are preventable. After all, how could a child accidentally drink furniture polish unless it was accessible to the child?

Parents may protest that it is not possible to watch toddlers at every moment of the day, especially when there are several young children in the family. Although often this is true, the key is to establish good habits for the use and storage of the many and varied household products, garden products, medicines, and the like in the home. All adults should inform themselves about the factors that contribute to accidental ingestions among children, and they should take steps to make sure these factors do not become operative in their homes.

The most obvious factor is, of course, accessibility. For many decades, poison prevention educators have been urging the public to read and heed the labels of all products brought into homes and to keep every potentially dangerous chemical out of the reach of children. Poison prevention legislation enacted in 1970 required certain potentially harmful products to be packaged in child-proof containers. Aspirin—which prior to the 1970s accounted for 20 to 25 percent of all deaths from accidental poisoning among children—was one of the first products required to be so packaged. Within a few years after this requirement became effective, the fatality rate from aspirin ingestion by children was reduced by more than half, demonstrating that legislation was more effective than all the education efforts.

Several other factors that play a role in accidental ingestions among children have been identified by the Los Angeles Medical Association Poison Control Center (PCC), which was started in 1957 as the Thomas J. Fleming

Memorial Poison Information Center. Year-by-year data on the causes and demographics of poisoning events reported in the United States is available on the Web site of the American Association of Poison Control Centers (www.aapcc.org).

It became evident to PCC personnel that many of the episodes had certain common features. One related to age. It had long been known that age was a factor in childhood poisonings, with almost all such cases occurring in children under the age of five. Data from the Los Angeles PCC refined the age factor: Essentially no accidental ingestions occur in children below the crawling age. The number increases until it peaks between the ages of two and three. After three years of age, the number drops rapidly so that by the age of five, a child is unlikely to become an accidental poisoning statistic. Any parent could explain these data; the child at greatest risk is one who has reached the age of sufficient coordination to find, hold, and pour a container but not the age of sufficient experience and knowledge to understand that all things are not edible.

A second factor, not recognized before its proposal by the Los Angeles PCC, relates to time of day. Poisoning episodes involving children occur primarily during the daylight hours, within which there are two peaks: one from 10 A.M. to noon and the other from 4 P.M. to 6 P.M. These are the hours just before lunch and dinner, respectively. Hungry children are more apt to eat anything they can put their hands on, and, if the parent is busy with meal preparation, less attention is being paid to what the child is doing.

A third observation by Los Angeles PCC personnel is that some toddlers are what might be labeled "repeat performers," whereas other children, sometimes even in the same family, just never become victims of accidental ingestion of chemicals. Repeaters seem to be very active, bright youngsters who are forever getting into some kind of mischief.

A parent who keeps all chemicals out of a child's reach and is especially alert to what a child in the critical age group is doing at all times, especially during the hours before lunch or dinner, is one who probably will never suffer the panic of a poisoning emergency. Of greater importance, the child of such a parent probably will never suffer the fright of a trip to an emergency room or the pain and discomfort of a stomach-pumping procedure.

Adults can also become victims of accidental intoxication, although their numbers are far fewer than those involving children. Adult accidents are usually the result of carelessness or ignorance on the part of others. Probably the most common cause of accidental poisoning in adults is the storage of pesticide solutions, solvents, household cleaners, or other chemicals in soft drink bottles or other kinds of food containers. Unsuspecting victims drink or eat from containers they believe contain the food specified on the label.

Such containers usually are left in easy reach on a table or shelf somewhere in the home. In some instances, unsuspecting victims have found containers used for storing excess pesticides or cleaning solutions in refrigerators.

Alcohol consumption leading to serious acute toxicity or death occurs primarily in college-age people, frequently as a result of hazing rituals or party binge drinking. Unfortunately, although these episodes get wide publicity and high schools and universities warn students about the problems of binge drinking, the serious consequences of consuming large quantities of alcohol over a short time have not been impressed on this population.

On a much larger scale, accidental spills of chemicals during transportation, either by railroad or truck, and massive accidental releases from industrial processes have the potential for acutely poisoning large numbers of people who have the misfortune to be in the vicinity when the accident occurs. In July 2009, a woman in South Carolina died and several others were hospitalized after driving into a cloud of ammonia that had leaked from a tanker truck and drifted over a highway.

The accidental release of an industrial chemical in an occupational setting can also result in multiple poisonings; however, such releases are often much smaller in scale and usually affect only one or a few people. Industry has a responsibility to anticipate likely accidents, take steps to prevent them, and make plans for emergency response in the event that an accident does occur. In addition, people who work with chemicals have a right to know about the toxicity of those chemicals so that they can take proper precautions to protect themselves during normal work procedures.

Unfortunately, since September 11, 2001, our sensitivity has been heightened to the risk of mass poisonings as terrorist acts. Chemicals such as sarin, a nerve agent that was used in the Iraq/Iran war and was released in the Tokyo subway in 1995 resulting in 12 deaths, and ricin, an extract of the castor bean plant (see Figure 6-3), have become everyday words in our society. These chemicals, as well as similar ones manufactured during World Wars I and II, have been outlawed by international convention since 1993; however, they are not terribly difficult to manufacture and, therefore, are relatively easy for governments or terrorist groups to obtain. Those who wish to delve deeper into these chemicals and their effects should review *Chemical Warfare Agents* (2007) by J. A. Romano and others, an excellent book that details weapons of mass destruction.

Society's concern about mass poisonings predates 9/11. In 1982, the Tylenol killings produced a scare that led to the secure closure packaging that is required today for over-the-counter medicines and some consumer products, such as deodorant and toothpaste. Seven people died shortly after taking Tylenol from bottles that were tampered with and contained potassium

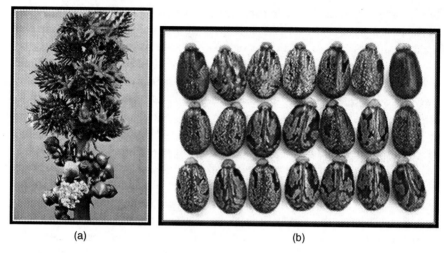

(a)   (b)

FIGURE 6-3   (a) Castor bean plant; (b) castor bean seeds from which ricin is extracted. (Copyright ©W. P. Armstrong, 2000.)

cyanide instead of the expected analgesic. As of 2010, no one had been convicted of these murders, and the reason for these killings is still unknown.

Antidotes

There is a common misconception in the public's mind that when a person has ingested a toxic substance and is rushed to an emergency room, the attending physician administers an antidote which, if given in time, reverses the course of what otherwise would be a fatal outcome. Unfortunately, such is not usually the case. An antidote is a chemical that by one mechanism or another counteracts the action of another chemical, thereby preventing it from exerting its toxic action. Many people believe that for every poison there is an antipoison (antidote). Actually, there are very few antidotal chemicals. In the great majority of cases of acute poisoning, all a physician can do is to try to remove as much of the toxic material as possible from the victim's system, treat whatever symptoms may be present, and support the life functions of the individual until the body's own restorative powers take over. This course of action, which is usually the only one available to emergency room personnel, is referred to as symptomatic and supportive treatment. Further, the few antidotes that are available may in themselves be toxic. Their administration does not necessarily ensure success. Too-vigorous treatment with an antidote also can be lethal.

Some people who are concerned about poisoning by pesticides have demanded that no chemical be registered unless it has an antidote. This demand is rather visionary, albeit noble and desirable, because of the paucity of antidotal chemicals for most classes of compounds. Legislation prohibiting the marketing of any chemical that does not have an antidote would not affect the class of chemicals we call pesticides as adversely as it would other chemicals that we use every day—drugs, household products, cosmetics, hobby products, automotive products, paints, varnishes, and so on. This is because, fortuitously, the majority of pesticides actually do have antidotes. A large number of pesticides are organophosphate compounds for which atropine or the chemical known as 2-PAM are antidotal. Many rodenticides are coumarin compounds that exert their toxic action by interfering with blood coagulation. Vitamin K is an antidote for these compounds. Arsenic insecticides have an antidote known as BAL. Acetic acid (the acid in vinegar) and its salts are antidotal for the very highly toxic rodenticide sodium fluoroacetate. Carbamates, another large group of pesticides, have atropine as an antidote. Many known antidotes are listed in Table 6-2.

It is interesting that nitrite, which has the antidote methylene blue for its toxic effects, also can serve as an antidote for cyanide; ethanol, which can

TABLE 6-2: Some Toxic Substances and Their Antidotes

Chemical	Antidote
Acetaminophen (Tylenol)	N-acetylcysteine
Arsenic insecticides	BAL
Atropine	Physostigmine
Benzodiazepines	Flumazenil
Beta blockers and calcium channel blockers	Calcium gluconate or glucagon
Carbamate insecticides	Atropine
Coumarin and Warfarin	Vitamin K
Cyanide	Nitrite, amyl nitrite, sodium thiosulfate
Ethylene glycol	Ethanol, thiamine
Fluoride	Calcium
Iodine	Starch
Iron	Desferrioxamine
Isoniazid	Pyridoxine
Lead	EDTA
Magnesium	Calcium gluconate
Methanol	Ethanol
Nitrite	Methylene blue
Opiods (heroin, morphine)	Naloxone (Narcan)
Organophosphates	Atropine and pralidoxime or 2-PAM
Oxalate	Calcium
Snake venoms	Specific antitoxins
Sodium fluoroacetate	Acetic acid
Thallium	Prussian blue

be acutely toxic, is the antidote for methanol. In the latter case, it is better to get a controlled dose of ethanol than to suffer blindness, mental incompetence, or death from acute methanol poisoning.

Fortunately, mild acute intoxications, regardless of chemical involved, almost invariably produce no demonstrable permanent injury. Recovery is complete, and health is restored. This is also true for most cases of severe poisoning when the victim responds well during the first critical hours or days after poisoning. The human body has remarkable restorative powers. Exceptions, however, such as permanent blindness after severe methanol poisoning and nervous system damage from some potent hallucinogenic agents do occur. An attending physician is the most reliable source of information concerning the prognosis for complete recovery in cases of severe acute poisoning.

CHRONIC TOXICITY

Chronic (repeat dose) toxicity refers to the harmful systemic effects produced by long-term, low-level exposure to chemicals. Far less is known about the chronic toxicity of nontherapeutic chemicals than is known about their acute toxicity, not because the subject is of any less importance or interest but because chronic toxicity is much more complex and subtle in its manifestations. Many symptoms of mild chronic intoxication are slow to develop; thus, the connection between exposure and illness may be obscured. In some cases, the symptoms may mimic those of other chronic diseases and, as a result, may be difficult to distinguish from some ills to which people are naturally prey.

The bulk of information available in the scientific literature on the chronic toxicity of chemicals relates to oral toxicity. The reason for this is twofold: First, some of the earliest concerns about chronic adverse effects were directed toward chemicals used in production, processing, and preservation of foods; thus, it was logical to study their oral toxicity. The second reason, and probably the more compelling one, is that the easiest and most convenient method of exposing test animals to chemicals is by the oral route. However, the importance of the skin and lungs as routes of chronic exposure to chemicals, and the importance of employing the same route in animal studies as that which is the usual route for human exposure, necessitate that chemicals be studied by these alternate routes. For chemicals of medicinal classes, chronic toxicity by alternate routes has been explored more fully.

The methods of study of chronic toxicity are described in Chapter 5. Data obtained from studies of chronic toxicity in animals are plotted on a graph

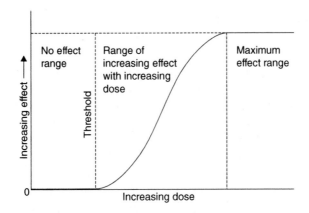

FIGURE 6-4   A dose-response curve.

to give a dose-response curve for the population of animals in the study like that shown in Figure 6-4. A dose-response curve has three parts. The first part of the curve is flat. It shows the range of doses that produces no detectable effect. The second segment of the curve begins at a point called the threshold and increases with increased doses until the maximum effect occurs. The third part of the curve is again flat. Effects described by the dose-response curve may be adverse, such as in toxicity studies, or beneficial (therapeutic), such as in pharmacological studies.

The curve is interpreted in this way: With chronic exposures of doses from zero up to the threshold, no negative effect is detectable because some biochemical or physiologic mechanism, such as metabolism, handles the chemical in a manner that prevents an effect from occurring. At the threshold, more compound is present than can be removed instantaneously, and the body is overwhelmed in some manner and effects begin to appear. This second part of the dose-response curve, from the threshold to maximum effect, describes a fundamental principle of pharmacology and toxicology: The degree of effect increases with increased dose, and increasing numbers of animals, including people, show the effect until finally a dose is reached where all of the animals show the effect. This third section of the curve is flat because the maximum effect has been achieved; no greater effect occurs with an increased dose.

The same curve can be drawn to describe the responses, either chronic or acute, of an individual to increasing doses of a chemical. Below the threshold, no effect is seen. Above the threshold, effects appear; they increase in number and severity until finally a dose is reached that the individual can no longer tolerate, and death occurs. When a therapeutic chemical is given at the threshold, its beneficial effects occur; as doses increase, negative effects result at some point and continue in severity until adverse events or death occur.

The dose-response principle—the relationship between dose and effect—aids in the interpretation of data obtained from chronic toxicity testing in animals in the low-, medium-, and high-level exposure groups. When an adverse effect increases with increasing dose, the effect is most probably due to the test chemical, whether the increase is proportional to dose or not. When an effect occurs to the same degree at all doses, it is probably not due to the test chemical. When the effect occurs in the low-exposure group and to a lesser extent or not at all in the middle- or high-exposure groups, the effect is almost certainly *not* due to the test chemical. In the last two cases, further experimentation may be necessary to clarify the matter.

No-Effect Levels and Thresholds

How are the data obtained from chronic toxicologic investigations used to make judgments about their significance for human health? Unfortunately, there is no nice, neat quantitative expression like the LD_{50} that can be applied to chronic toxicity. Instead, concepts such as no-effect levels, thresholds, and margins of safety must be relied on.

The no-effect level of a chemical is an experimentally determined quantity. It is a quantity of chemical to which laboratory animals are chronically exposed, expressed in parts per million or billion (ppm or ppb) in the diet or milligram per kilogram (mg/kg) of body weight or mg per square meter (mg/m^2) of body surface area, that produces no effect when compared with control animals. Obviously, the no-effect level in a given experiment will vary with the caliber of scrutiny to which the animals are subjected. The more gross the parameter being measured, the greater are the quantities of chemical required to produce a detectable difference between experimental and control animals.

The more subtle the parameter being measured, however, the smaller is the quantity of chemical required to produce the effect. For example, some of the older, classic liver function tests are much less sensitive indicators of liver damage than the current tests that measure the activity of certain liver enzymes. A considerable amount of liver damage must occur before it can be detected by liver function tests, whereas liver enzyme patterns can show change with very minimal amounts of damage. Thus, a quantity of liver toxin that would be a no-effect level when measured by the older liver function tests could be an effect level when measured by liver enzyme activity. As a general rule, a no-effect level is taken to be that dosage of chemical to which experimental animals are chronically exposed that produces no harmful effect detectable by toxicologic techniques that are generally accepted by the scientific community as being current and appropriate.

The term *threshold* is used to describe the dividing line between the adverse no-effect and adverse effect levels of exposure. It is the maximum quantity of a chemical that produces no negative effect or the minimum quantity that does produce an effect.

The threshold for a given effect can, and usually does, vary with species, with individuals within a species, and perhaps even with time in the same individual. A threshold is therefore an elusive quantity that is impossible to determine precisely and directly by experiment. Even if the threshold of an effect in a given individual could be ascertained, it would apply only to that individual. Despite the fact that thresholds in individuals cannot be determined precisely, the existence of thresholds is generally accepted as fact. The threshold concept is of great importance to the understanding of the toxic action of chemicals.

Since low levels of exposure produce no detectable effects and high levels produce an effect, and the dose-response relationship is continuous, a threshold must exist. For purposes of extrapolating animal data to humans, the highest level of exposure that produces no detectable adverse effect of any kind in any of the test animals is used by toxicologists as the threshold.

Margins of Safety

The uncertainties inherent in extrapolating data from animals to humans require that margins of safety be used in the process. A *margin of safety* is an arbitrarily established separation between the highest level of a chemical that produces no adverse effect in any animal species and the level of exposure estimated to be safe (or effective for its use in the case of drugs) for humans. The FDA adopted the convention of a 100-fold margin of safety many decades ago when it first began setting standards for legally acceptable quantities of food additives, such as colors and preservatives, that could be added to processed foods.

The assumptions behind the 100-fold margin are that (1) humans are 10 times more sensitive to the adverse effects of chemicals than test animals are, and (2) the weak in the human population (young, old, debilitated, etc.) are 10 times more sensitive than healthy adult humans. Multiplying 10 times 10 gives the 100-fold margin. Thus, for example, if the highest level of a chemical that showed no adverse effects in any animal species tested was 100 ppm in the diet, the maximum dietary concentration of that chemical that could be considered safe for humans would be 1 ppm.

This is a much oversimplified example of the kind of calculation that is made in extrapolating data from animal studies to humans. The 100-fold margin of safety is not set in concrete. In fact, it has been facetiously suggested that if humans had been created with 8 fingers instead of 10, the classic margin of safety probably would have been set at 64 (8 times 8). Many other factors—such as the toxicities of the chemical by skin absorption and inhalation relative to its toxicity by ingestion, the potential for exposure by routes other than oral, the kinds of foods that will contain the chemical and the relative contribution of each food to the total dietary intake, the need for the use of the chemical, the existence of other less toxic chemicals that could serve the same purpose, evidence of mutagenicity, teratogenicity, or carcinogenicity, and data from human experience with the chemical—are all taken into consideration in the official deliberations and hearings that precede the establishment of standards for chemicals that may legally be added to human or animal foods.

In recent years, there has been a call for agencies regulating environmental chemicals to adopt a 1,000-fold margin of safety, rather than 100-fold, in their standard-setting procedures. Proponents claim that an extra 10-fold margin of safety is required when toxicity data are considered to be insufficient, which is the usual case. Perhaps the 1,000-fold margin of safety is just an extension of the logic that says if a little is good, more is better: If a little is safe, less is safer.

Actually, there is no more scientific justification for a 1,000-fold margin than there is for a 100-fold margin. If an exposure standard appears to be reasonably protective, reducing it by a factor of 2, 10, or 1,000 cannot make it more safe. The important point is not the margin but that all factors related to use of the chemical are taken into consideration and that a permissible exposure level is set within the no-effect range. For some chemicals, a small margin of safety is completely protective, whereas for others, a larger margin is required.

For medicinal chemicals, the acceptable safety margin depends on the type of disease that is being treated. For example, a safety margin of 1 (or less) may be appropriate for a cancer chemotherapeutic where the patient has a high probability of death if not treated. We all have heard of how toxic chemotherapy is, and this toxicity is accepted in the circumstances of a lethal disease. The safety margin should be considerably higher (perhaps 25-fold) for a drug to treat eczema in children. The higher safety margin is set because the disease in not fatal and the population is young, so society is unwilling to risk a treatment that has substantial toxicity. Note that the safety margins for drugs are considerably smaller than those for environmental chemicals. There are two main reasons for this:

1. Drugs are therapeutic, and there is a *benefit* derived from their use. (See Chapter 11 for a full discussion of benefit/risk analysis.)
2. People knowingly take medicine, and we as a society are more willing to assume voluntary risks rather than risks from inadvertent exposure.

The threshold concept and margins of safety are of great importance to toxicologists because they allow for judgments about the potential hazard, or lack thereof, to humans from long-term exposure to very small quantities of foreign chemicals. However, these concepts do not permit toxicologists to answer other questions that are most often asked by public officials, legislators, legislative aides, and regulatory staff people as well as by private citizens. The majority of such questions concern how large a chronic exposure to a specific chemical can be tolerated by individuals or populations without any ill effect. It is difficult, even for people with some scientific knowledge, to accept the fact that no one knows, or can know, what the maximum safe chronic exposure is (or the minimum harmful exposure) for all people for any particular chemical.

Sufficient Challenge

Although the effects produced by nontherapeutic chemicals are considered to be detrimental, this is not necessarily the case. Toxicologists who have been engaged for any period of time in research into chronic toxic effects of chemicals may have observed that animals in the group with the lowest exposure to a test chemical grow more rapidly, have better general appearance and coat quality, have fewer tumors, or live longer than the control animals. In a paper published in *Food and Cosmetics Toxicology* in 1967, the noted toxicologist Henry F. Smyth Jr. proposed the term *sufficient challenge* for this phenomenon. In a comprehensive work, *The Reverse Effect* (1988), documenting the contradictory effects (detrimental and beneficial) of a host of physical and chemical agents, Walter A. Heiby applied the more encompassing term *reverse effect* for this phenomenon.

The phenomenon of beneficial effects from exposures to trace quantities of foreign chemicals, although often a subject of conversation among toxicologists, particularly with regard to why such effects occur, is rarely mentioned in the scientific literature. When it occurs in a chronic toxicity experiment, the text of the paper reporting the results seldom mentions the fact. Only a careful perusal of the data tables and figures presented in the body of the text reveals the phenomenon. Such subtleties are lost on people who read only the abstracts of scientific papers.

Unfortunately, some scientists may be counted among the abstract-only readers.

The reluctance of toxicologists to acknowledge and discuss freely the observation that trace quantities of foreign chemicals can produce beneficial effects may be founded partly on the fact that they have not, as yet, formulated a unified theory for the phenomenon. Or it may be because we live in a time when it is not politic to make favorable statements about synthetic chemicals. Toxicologists who work for industry are particularly hesitant to discuss the phenomenon publicly. No matter how scrupulous an industry scientist may be in maintaining objectivity in scientific evaluations and judgments, any statements he or she makes may be rejected by some people as being biased.

Since toxicologists recognize that the phenomenon has little practical significance (unless it could be put to some therapeutic use), they have no compelling reason to emphasize it; they might be misunderstood and, as a result, lose a reputation for objectivity. In the absence of any public benefit, why open oneself to attack from people ready to seize on any statement that sounds prochemical and label its author as an industry apologist? Dr. Smyth learned firsthand how easy it is to be misunderstood; his first use of the term *sufficient challenge* earned him the epithet "Dr. Smyth and his fellow poisoners."

However, it should not be surprising that synthetic chemicals can convey some therapeutic use, since most medicinal chemicals are synthetic and have been synthesized specifically for therapy. Nontherapeutic chemicals that show low-dose effects surely work through the same biological pathways, such as binding to receptors, as do therapeutic chemicals.

Figure 6-5 represents how the concept of sufficient challenge changes the shape of the dose-response curve. With small doses, no effect on the health

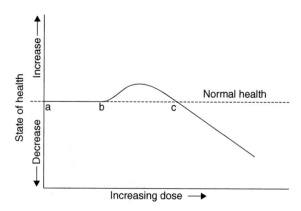

FIGURE 6-5   Dose-response curve for a compound expressing sufficient challenge (hormesis).

of the experimental animals is observed (points a to b). Animals receiving slightly higher doses exhibit better-than-normal health (points b to c, the range of sufficient challenge or therapeutic effect in the case of a drug). At point c, the curve passes back through normal health and, from that point on, deleterious effects occur.

A general acceptance of the concept of sufficient challenge would have no impact on chronic exposure standards for environmental chemicals or the regulatory procedures that govern them. The prohibitive cost of determining the dose ranges of beneficial effects for nontherapeutic chemicals, the impossibility of determining individual responses to such doses, and public concern about environmental compounds in general all operate against any practical application of the concept. The concept could, however, help bring a very necessary public objectivity to the subject of environmental chemicals, similar to that in place for medicines.

A phenomenon referred to as *hormesis* has gained acceptance among several groups of scientists. In 2009, Dr. Edward Calabrese of the University of Massachusetts at Amherst received the Marie Curie Prize from the International Dose-Response Society for his work on hormesis, low-dose radiation, and health. He had performed multiple experiments on low doses of minerals such as chromium and selenium and proposed that the beneficial effects of small doses of certain compounds result from the fact that they act as a toxin to simulate biological systems in a positive way, even though higher amounts are overtly toxic. (See Chapter 4 for a discussion of the toxicity of large doses of selenium.)

Bioaccumulation

No study of the chronic toxic effects of chemicals would be complete without a discussion of bioaccumulation. *Bioaccumulation* is a term that is commonly used but apparently has different connotations for different people. For some it means the accumulation of cellular or tissue injuries as a result of exposure to chemicals. It is more popularly used to refer to an increase in the concentration of a chemical in specific organs or tissues over that which would normally be expected. The latter meaning will be used in the discussion here. It should be noted that the terms *storage* and *bioaccumulation* are not synonymous. *Storage* means deposition of a chemical in some anatomic site. *Bioaccumulation* is the process by which the quantity of a chemical is increased in an anatomic site.

The phenomenon of bioaccumulation is poorly understood and is therefore frightening. The public became aware of the concept decades ago when environmental concerns about chlorinated hydrocarbon pesticides, particularly DDT, were being widely publicized and today when lead accumulation is discussed. However, people still may be confused and concerned about what bioaccumulation is and what it means for their health and well-being.

A brief description of the nature of the chemical balances in living organisms is essential for an understanding of bioaccumulation. All living cells possess a certain specific composition, which, within normal limits, remains constant as long as they maintain their customary good health. Although this relatively stable composition makes it appear that living organisms are static organizations, nothing could be farther from the truth. All creatures live in what is known as a state of dynamic equilibrium. *Dynamic* means "moving," and *equilibrium* refers to the balance, or equalness, that is maintained within living organisms.

For a simple illustration of a state of dynamic equilibrium, consider a box that contains a specific number of marbles; every time some new marbles are added to the box, an equal number of old marbles is removed. The total number of marbles in the box remains constant, but the individual marbles present in the box may change through time. If the input of marbles increases, the output will also increase by the same number. If there is no delay between input and output, the total number in the box remains constant. If there is a delay in the output, the total number in the box will increase as the new marbles are added. The magnitude of the increase will be determined by the length of the delay between the increased input and the increased output. So it is with living organisms: There is a steady drive toward the achievement of equilibrium—constant change, but little or no net change. Even the bones in our bodies, which people think of as inert structures, contain atoms and molecules that are in a constant state of flux.

With the concept of dynamic equilibrium in mind, let us now look at bioaccumulation. Whether a chemical accumulates in an organism depends on how fast it is eliminated (metabolized or excreted) relative to how fast it is absorbed into the body. The time between absorption and elimination is the *residence time* of the chemical. The term *biological half-life* is used to express the length of time a chemical resides in the body. Biological half-life is a concept similar to radioactive half-life, where the half-life represents the time it takes for half the molecules to decay to a nonradioactive isotope; biological half-life is the time required for the quantity of the chemical in the body to be reduced by half.

For chemicals with the same rate of absorption, those with longer residence times have a greater potential for bioaccumulation than those with shorter residence times. If the residence time is very short, little or no bioaccumulation will occur (at least not of the chemical, itself; however, a metabolite may accumulate, but since most metabolites are less toxic than the parent compound, we will ignore this phenomenon here). Residence time is dependent on such factors as pathway through the body, rate of metabolism or excretion, and the tendency of the chemical to be retained for a period of time in some metabolic pool or storage site (depot), such as fat or bone. The actual concentration of a chemical in a living organism is regulated by the same kinds of forces that drive the organism's internal biochemical environment toward equilibrium.

The storage of foreign chemicals in various depot sites can involve complex physiologic processes and equilibrium relationships. Although, for the sake of simplification, the next discussion considers only the overall equilibrium among absorption, storage, and elimination, please remember that the blood, which serves as the transport medium for chemicals in the body, participates in the process. Blood has its own equilibrium relationships at the sites of entrance, storage, and elimination of chemicals.

The simplest case involving storage of chemicals occurs when the concentration of chemical to which an organism is exposed remains constant over a prolonged period of time and, after entering the body, is not metabolized. Before the onset of exposure, there are no molecules of the chemical in the body's storage site. When exposure starts, molecules of the chemical begin entering the depot. Some of these molecules also exit the depot, but, because of the time delay between entrance and exit, the number entering is greater than the number exiting. The result is that the concentration of chemical in the depot site gradually increases. With continued exposure, the concentration of chemical in the storage depot continues to increase until finally it comes into equilibrium with the exposure concentration.

At this point, the number of molecules leaving the storage site equals the number entering, and the storage concentration remains constant. Molecules of the chemical are moving constantly in and out, but there is no net change—just like the marbles in the box. If the exposure concentration increases, the storage concentration will gradually increase until it reaches a new equilibrium with the higher exposure; likewise with decreased exposure, it will gradually decrease until it comes into a new equilibrium. If exposure ceases, the stored chemical will gradually be eliminated from the body.

The quantity of chemical stored in any body depot can never exceed that which would be in equilibrium with the exposure. The chemical cannot remain in the storage depot without being replenished continually from the outside. Thus, the popular notion that foreign chemicals stored in a depot become immobilized and permanently fixed in the body, with additional exposures increasing the quantity stored *ad infinitum,* has no basis in fact. The claim that our bodies can become "walking time bombs" is untrue.

The quantity of storage and the length of time required to reach an equilibrium state varies from chemical to chemical and depends on the magnitude of the exposure concentration and the nature of the chemical— that is, its physical, chemical, and biologic properties. Many chemicals have no affinity for any storage depot in the body and thus do not store. A few have a large propensity for storage and with high exposures can build up to high concentrations. The remainder display varying degrees of storage between the two extremes. If exposure to a chemical varies continuously in concentration and duration, interspersed with periods of no exposure—as is the case with many of our trace exposures to foreign chemicals in the environment—an equilibrium state is never achieved. In such cases, the quantity stored in the body is also continually changing but would be considerably less than that which would be in equilibrium with the highest exposure level.

The relationship between magnitude of exposure and storage concentration at equilibrium is not known for most non-therapeutic chemicals. One exception is the organochlorine pesticide DDT, the subject of a tremendous amount of investigation, including study of its storage and excretion patterns. Before the use of DDT was banned, the U.S. population on average received a daily oral exposure of approximately 0.2 ppm DDT in the total diet during the several decades of heavy use of the pesticide. This level in the diet produced an average DDT concentration of 7 ppm in the body fat of the American people. The increase in storage concentration relative to exposure concentration at this low level of exposure was approximately 35-fold.

In studies with rats and mice, a diet containing 20 ppm DDT produced a body fat content of 200 ppm, a 10-fold increase. A diet of 200 ppm produced body fat levels of 600 ppm, only a 3-fold increase. Thus, for DDT, the relationship between exposure and storage concentrations is not linear. The total quantity of storage increases with increasing exposure, but low levels of exposure produce relatively larger storage levels than do high levels of exposure. From what is known about storage and excretion of foreign chemicals, it appears that DDT can serve as a model for chemicals that store in body fat.

For therapeutic drugs, bioaccumulation must be taken into account and is determined during the development of the compound. Because people find it easier to remember to take a pill each day, daily dosing is a preferred method of administering drugs. Frequently, a medicine will not reach the level needed for efficacy until several days after dosing. If earlier therapeutic levels are required, a *loading dose* is administered. For example, the drug digoxin, originally extracted from the foxglove plant but now synthesized, is used to treat various heart conditions. Because it may take five or more days to reach the therapeutic level, the doctor administers a loading dose on Day 1 that will bring the patient's blood levels into the therapeutic range right away—it is not a good idea to wait five days when you have a serious heart problem. The patient is then given a maintenance daily dose to keep his or her blood levels at that therapeutic level, thus maintaining equilibrium by replacing the metabolized and excreted digoxin with just enough to prevent further bioaccumulation. The caveat with digoxin is that the toxic level is quite close to the therapeutic level (the drug has a narrow therapeutic index), and there is considerable biodiversity from patient to patient; therefore, it is a difficult task to know what dose to load and what dose to maintain. This is why the doctor will order digoxin blood level tests frequently during the early weeks of therapy in order to monitor closely each individual's response to the drug. Further complicating the digoxin story is that the toxicity of digoxin mimics the efficacy—that is, both toxicity and therapy have the heart as the target organ. This drug, which has been used for over 100 years, is responsible for many hospitalizations each year due to its side effects.

Another piece of misinformation that has been circulated about bioaccumulation is that it puts every one of us at risk of serious or fatal poisoning. It is claimed that chemicals suddenly released from fat depots in the event of a severe weight loss, such as could occur in the case of debilitating illness, serious dieting, or starvation, can cause acute poisoning. Poisoning from rapid mobilization of a chemical from fat stores is theoretically possible in laboratory animals under rigidly controlled conditions of exposure and food intake but is a practical impossibility in the human population.

The question of whether DDT could be released from body fat with sufficient rapidity and in sufficient quantity to produce symptoms of poisoning was of interest to FDA scientists during World War II, when heavy applications of DDT were employed to prevent pandemic outbreaks of insectborne diseases such as malaria. In FDA studies, rats were fed 200, 400, or 600 ppm DDT in their diets for periods sufficient to build up high levels of DDT in the body fat, followed by total withdrawal of feed. Rats receiving 600 ppm suffered marked tremors typical of acute DDT intoxication; rats receiving 200 or 400 ppm DDT suffered increased irritability but no tremors.

How do these quantities relate to acutely toxic doses for humans? It is known from controlled studies in human volunteers that ingestion of 35 mg DDT/kg of body weight per day, for a period of four to five years, produced no observed adverse effects, acute or chronic, in any of the subjects. It is also known from accidents and volunteer experiments that ingestion of somewhere between 50 and 70 mg/kg is less than an acutely toxic dose. We can conclude that people exposed to moderate levels of DDT daily do not have a "DDT dump" that puts them at risk for the acute signs of toxicity.

Bioaccumulation is not inherently good or bad, but in the public mind, it is considered the latter. In actual fact, whether it is beneficial or detrimental depends on the context in which it occurs. The ability of an organism to store a chemical in some anatomic depot can be beneficial when it functions as a mechanism to protect against the toxic action of a chemical, and it does no harm while it resides in a storage site. For example, lead is a very toxic element for animals. It stores in bone, where, like DDT in fat, it is in equilibrium with blood levels, which in turn are in equilibrium with exposure and excretion. Lead isolated in bone does no harm, but lead in nerve tissue causes very serious damage. The removal of lead from blood by bone may prevent the blood lead level from becoming sufficiently high to damage nerve tissue. If lead exposure is prolonged or excessive, the bone storage site

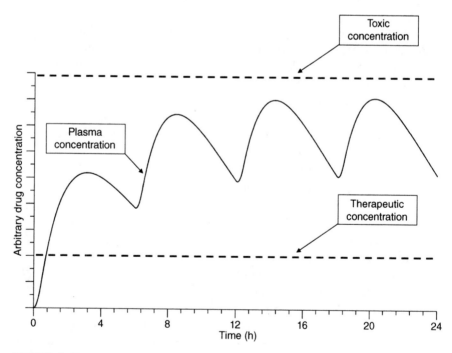

FIGURE 6-6 Theoretical plasma concentrations after four doses in a 24-hour period. (*Source:* From Eric Grovender, Medtronic.)

becomes saturated and can no longer remove lead from blood. In this case, blood lead levels can reach levels toxic to nerve tissue.

A storage depot may have a buffer function during periods of increased exposure. The chemical deposited in the storage site is prevented from exerting a harmful effect rather than building up in the blood to a level where it will damage a sensitive organ. When exposure decreases, the chemical moves from the depot and is eliminated from the body, thereby freeing the site for some future time when exposure may again increase and storage is required to protect against adverse effects somewhere else in the body. When exposure levels are sufficiently high to saturate the depot, the protective effects of the storage site ends. The process of bioaccumulation allows therapeutic chemicals (medicines) to be administered sporadically (e.g., once, twice, three, or even four times per day) rather than continuously. Figure 6-6 shows how sporadic administration allows for blood concentrations to remain above the therapeutic level but below the toxic level, providing efficacy for the patient without pushing concentrations into a toxic range. The number of doses per day is determined by the length of the plasma half-life of the drug: The longer the half-life, the fewer doses per time interval that are needed. If a single large dose would be given, the time spent in the therapeutic range might be the same as with multiple smaller doses, but the maximum concentration would be above the toxic level and then would fall below the therapeutic level before the next dose was administered. Because the half-life is a property not only of the drug but also of the metabolism of the patient, refining the doses can be difficult.

CHAPTER 7

MUTAGENESIS AND CARCINOGENESIS

MUTAGENESIS

The word *mutagenesis*, the production of a change in the genetic material of an organism, is derived from Greek and Latin words. *Gennan* is the Greek word meaning "to produce," and *mutare* is the Latin word meaning "to change." Mutagens are physical or chemical agents that produce genetic changes (mutations). Mutations are considered to be the initiating events in the development of cancer and in some birth defects. Before considering these topics, we need to have some understanding of mutagenesis and the structures affected by mutation.

Genetic Code

In every living cell, whether it is a microorganism, plant, or animal, there is a system of coded messages that tells it how to make more cells exactly like itself. This message system, known as the genetic code, is contained in complex structures called *chromosomes*. Chromosomes are composed of molecules of deoxyribonucleic acid (DNA) and are located in the nuclei of cells. Ribonucleic acid molecules (RNA) in the cytoplasm of cells also participate in the message system. DNA and RNA molecules are composed of subunits called *nucleotides*. Each nucleotide is composed of 3 molecules chemically bound together: a purine or pyrimidine base, a pentose sugar, and phosphoric acid. DNA and RNA nucleotides differ from each other in the pentose sugar they contain; DNA contains deoxyribose, and RNA contains ribose. Almost all nucleotides contain one of only 2 purine or 2 pyrimidine bases; therefore, there are only four principal kinds of nucleotides.

DNA nucleotides join together to form larger units called *genes*. Genes are considered to be the smallest units that carry a genetic message. Genes in turn are linked together in long chains that twist around themselves in

The Dose Makes the Poison: A Plain Language Guide to Toxicology, Third Edition.
By Patricia Frank and M. Alice Ottoboni
© 2011 Patricia Frank and M. Alice Ottoboni. Published 2011 by John Wiley & Sons, Inc.

the shape of a double helix. A long double helix strand of DNA makes up a chromosome. James Watson and Francis Crick won a Nobel Prize in 1962 for the discovery of the helical structure of DNA, which they explicated from data collected by Rosalind Franklin.

Although the number of nucleotides is small, a tremendous number and variety of genes can be formed by linking various nucleotides together in different sequences. Three nucleotides linked form an amino acid, and many amino acids linked together form a gene. Genes also can be linked together in many different sequences to form a wide variety of chromosomes. The nucleotides may be likened to letters of the alphabet and genes to words formed from the letters. Letters and words can be combined in a great variety of ways to form sentences that can convey an infinite number of meanings. So it is with chromosomes.

Work by the publicly funded Human Genome Project and Celera Genomics, a privately funded group, has led to giant steps forward in understanding the composition of the genetic code for humans as well as for many other species. In a fierce competition between these two institutions, the sequence of the human genetic code of about 25,000 genes was completed in the unbelievably short time of 13 years. The race to sequence other living forms continues. Almost every month a press release announces the complete sequencing of another species. Tremendous steps in this area were made possible by the refinement of sequencing technology, including robotic instrumentation that has speeded up the whole process. However, gaps remain in our knowledge of how the genetic code leads to a fully formed adult animal or plant.

The number of chromosome pairs varies with species, but all members of the same species and all cells of individuals within that species contain the same number. For example, fruit flies have 8 chromosomes (4 pairs) and humans have 46 (23 pairs). The known genetic diseases associated with each pair of chromosomes are extensive and ever growing. Determination of the genetics of a disease assists in designing therapies to ameliorate or even cure such diseases and hence has attracted the attention of pharmaceutical companies.

Mutations

A mutation is a change in the genetic code of a cell that results in some change in its message. If the mutated cell is capable of reproducing itself, it will divide into two cells that will be made according to the new instructions, and thus they will carry and reproduce the new code. These two cells and their progeny will all be different in some way from the parent cell, and the changes will depend on the dictates of the new code. The change may be so small that the new cells will seem to be exactly like the parent cell in form and function, or it may be so great that the new cells will be very different.

Mutations may occur in somatic cells (those not related to reproduction), or they may occur in germ (reproductive) cells. Somatic mutations are not inherited by offspring of the parent in which the somatic mutation resides. For example, a mutation to a skin cell will not be passed on to an offspring. If the change is small and the mutation is uncorrected, the somatic cells derived from the mutated cell probably would survive and reproduce and most probably would never be discovered because they would be few compared to the many, many normal cells surrounding them. If the change is a major one, the mutated cell may die before dividing. If it divides, the daughter cells eventually may die if they are not compatible or competitive with their normal neighbor cells. If they are compatible and competitive, they may grow to form a tumor. The tumor may be either benign or malignant (cancerous).

A mutation that occurs in a germ cell (sperm or egg) can be inherited. As with somatic cells, a change in a germ cell may be small and compatible with life, or it may be severe and kill the cell. If the cell survives and a new individual is formed from it, all of the cells of the new individual, both somatic and germ cells, will contain the change. If the new individual is successful in reproducing, the mutation has the potential to be passed on to all future generations and become part of the gene pool of its species (see Chapter 8).

Mutations may be dominant or recessive, a very important distinction in germ cells. When an individual forms mature sperm or eggs, each cell receives one strand of DNA from each pair of chromosomes. As a result, each germ cell carries one-half the complement of chromosomes that is normal for its species. In humans, germ cells contain 23 chromosomes. When a sperm and an egg unite, the two homologous chromosomes find each other and form a new pair. A recessive gene on one chromosome of a pair is overridden by the dominant gene on the companion chromosome. In this case, a mutated recessive gene does not express itself but is inherited and can be seen in some future generation if it combines with a companion gene that is also recessive. A dominant gene overrules a recessive gene and is seen immediately. Thus, a mutated dominant gene will overcome a recessive gene, whether the latter is mutated or not.

Some genes are neither dominant nor recessive, but rather a blend of the two, as demonstrated by the classic experiments of the geneticist Gregor Mendel with four o'clock flowers (a plant so named because the flowers open late in the day). White four o'clocks have a pair of genes for white

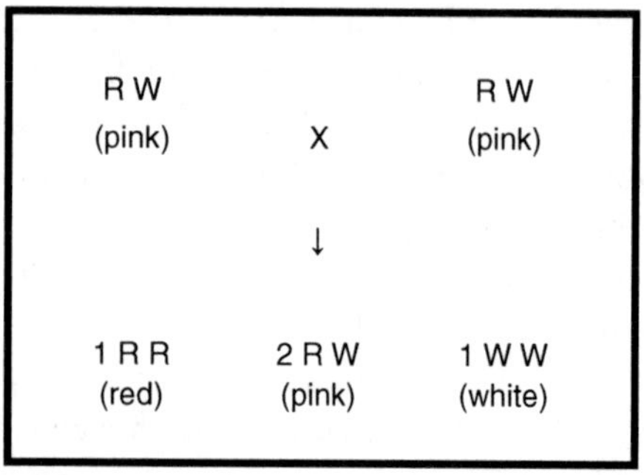

```
        R W                    R W
       (pink)        X        (pink)

                     ↓


       1 R R      2 R W      1 W W
       (red)      (pink)     (white)
```

FIGURE 7-1   An example of pure Mendelian inheritance where R = red, W = white and P = pink.

(WW), and red four o'clocks have a pair of genes for red (RR). Thus, the germ cells of red flowers contain one R gene, and the germ cells of white flowers contain one W gene. When white and red flowers are crossed, their genes combine to yield all pink-flowered plants (RW), a blending of red and white. When pink-flowered plants are crossed, the red gene from one can combine with either a red or a white gene from the other, yielding a red-flowered plant (RR) or a pink one (RW), respectively. In the same manner, the white gene can combine with a red or a white gene from the other, yielding a pink (RW) or a white (WW) plant, respectively. Thus, when pink flowers are crossed, their R and W genes recombine to produce red, pink, and white flowers in a ratio of 1: 2 1 (see Figure 7-1)

Color in four o'clock flowers is a very simple example of blending when neither gene of a gene pair is dominant. Many physical characteristics are dictated by more than one gene pair. These pairs may display dominant traits, recessive traits, or blending combinations of these two, greatly complicating the expression of a trait.

Significance of Mutations

Many people have the false impression that mutations are always harmful to an organism. Some mutations are harmful, but the majority is apparently of no consequence, and an important few are beneficial. Beneficial mutations may make individuals more disease resistant, better able to obtain food, or in some way better able to cope with a changing environment. As a result, an individual with a mutation may thrive and propagate while an individual without the mutation may be less successful in producing offspring.

Without mutagenesis, evolution cannot occur, since the formation of new species is the accumulation of small mutations over time or, in some cases, a large mutation that occurs quickly. It is interesting to speculate about the tremendous diversity of characteristics found in the human population and the role that mutations have played in variations, such as color of skin, eyes, or hair; stature; size and shape of eyes, noses, ears, and lips; and susceptibility and resistance to disease. For many characteristics, it is difficult to see how minor mutations could be anything other than neutral. However, mutations relating to skin color, for example, might be classified as harmful or beneficial depending on the geographic area. Prehistoric humans evolved in equatorial zones where sunlight was very intense. A light-skinned prehistoric human, without benefit of fur or clothing, would not survive very long because of severe sunburns and resulting infections and cancer, whereas dark skin would be protective against solar rays.

As prehistoric humans moved from the tropics to colder climates with less intense sunlight, dark skin would be a disadvantage because of its lesser ability to absorb the ultraviolet rays of the sun, which are required for the synthesis of vitamin D in the skin. In areas of weak sunshine, dark-skinned people would tend to become ill or die from vitamin D deficiency, whereas light-skinned people would survive and propagate. The influence of skin color on survival in diverse geographic zones is no longer of such importance since modern humans have clothing to protect against excessive sunshine and vitamin supplementation to protect against vitamin deficiencies.

An interesting example of a mutation that was beneficial when it occurred many, many generations ago but is detrimental in modern times is the sickle-cell trait, a change in the shape of red blood cells. Certain swampy marshlands provide perfect breeding grounds for mosquitoes that carry malaria. The disease was endemic, and many children died of malaria before reaching adulthood. But sickle-cell trait protected its carriers against malaria. As a result, people with the trait could survive to adulthood and reproduce. The number of individuals who died from the anemia that resulted from full expression of the trait was fewer than the number who would have died from malaria. Because sickle-cell is a recessive trait, full expression of the trait occurs only in offspring of parents who both are carriers of the trait and who each contribute a sickle-cell gene to their child.

Sickle-cell anemia is a debilitating disease that may be fatal by early adulthood if left untreated. In the United States, sickle-cell trait is undesirable because it provides no benefit and may cause harm; it is not needed to protect against malaria, a disease rare in the richer countries of the world and treatable when it occurs. However, malaria is still a problem in wide areas of the world where mosquito breeding conditions exist and the high cost of antimalaria medicine makes treatment unrealistic. The Bill and Melinda Gates Foundation has been active in funding new and affordable medicines to treat malaria in poor countries, and it is hoped that malaria will join the ranks of polio as a disease of the past. Additionally, the pharmaceutical company Novartis has been selling an effective drug to treat malaria to underdeveloped nations at less than cost so more people can be treated for less money. Meanwhile, the sickle-cell trait is harmful in modern society because people who are carriers of the trait run the risk of having a child with the disease. Such a child suffers physical debility and pain, and its parents suffer psychological trauma because of this suffering.

Global warming is affecting the world distribution of malaria and other tropical diseases. As areas outside the tropical zones around the equator warm, tropical diseases are spreading where favorable conditions for the incubation of disease vectors, such as mosquitoes, tsetse flies, and various parasites, can thrive. We are going to be fighting these infectious disease battles for the foreseeable future and in areas where these diseases may not have been seen before.

All organisms have biochemical repair mechanisms within them that protect against mutations. In fact, there are several kinds of repair mechanisms so that if one fails, another can be brought into play. In simple terms, the repair systems read the DNA code and when they find a mistake or a changed message, they correct the message to make it read as it did originally. Individuals vary in the quantity or activity of their mutation-corrector mechanisms. For example, people who have a hereditary deficiency of the mechanism that repairs the damage done to skin cell DNA by ultraviolet light may suffer from the disease known as xeroderma pigmentosa, a condition that predisposes them to skin cancer.

Mutations are not rare events. All of us normally carry within us a great many mutated cells. The causes, nature, and significance of this background incidence of mutations are under active investigation. Once major cause of spontaneous mutations is cosmic radiation and radiation from soil and rocks, both of which are natural and vary with geographic location. Publicity about the presence of radon, a naturally occurring radioactive gas, in home and office environments brought the issue of natural radiation into public focus. We can modify slightly our exposure to natural radiation by increasing ventilation in our homes and offices or by changing our places of residence, but we cannot escape it completely. We also need to realize that we are exposed to more natural radiation at higher altitudes and when flying. However, it is important to eliminate as much natural radiation from our lives as we can, so monitoring for radon in the basement is appropriate. Remedial action should be taken if the levels are found to be of concern. The Environmental Protection Agency (EPA) recommends testing for radon since it estimates that radon exposure causes about 21,000 lung cancer deaths in the United States each year. See the EPA Web site (www.epa.gov/radon/pubs/citguide.html) for information on testing for radon and how to reduce radon levels in your home.

In addition to our baseline exposure to naturally occurring radiation, our exposure to synthetic radiation is increasing. Medical tests such as computed tomography (CT) scans, positron emission tomography (PET) scans, bone density scans, and mammograms increasingly are being performed. According to *AARP Magazine* (May/June 2009), it is estimated that the radiation exposure from one bone scan is equivalent to about one day of naturally occurring radiation. A CT scan of the abdomen or spine, however, can be equivalent to up to three years' exposure to naturally occurring radiation, while a scan of the head might be eight months' worth. Improvements in mammography are reducing this exposure to radiation, making the mammogram among the lower exposure tests. Magnetic resonance imaging (MRI) and ultrasound scans do not produce radiation exposure.

Mutation and Cancer

Interest in how and why mutations occur has been stimulated to a large degree by the evidence that mutation plays an integral role in cancer development. Scientific investigation of mutagenesis produced by radiation predates that of chemicals by several decades. The development and use of nuclear weapons during World War II greatly increased scientific interest in radiation-induced cancer and its underlying cause, mutagenesis. However, despite the fact that the first cancer attributed to exposure to chemicals was described as long ago as 1775, when Sir Percivall Pott drew attention to exposure to soot as a cause of scrotal cancer in chimney sweeps, there was little interest in chemical mutagenesis until relatively recently, when it was rediscovered that some chemicals such as asbestos, benzene, and vinyl chloride could cause cancer.

In the early days of investigation into the mechanism whereby chemicals caused cancer, it was natural to look for clues from the wealth of data already available on cancer produced by X-radiation (X-rays). Since it was accepted that X-rays caused cancer by inducing mutations, it seemed logical to assume that chemicals also caused cancer by inducing mutations and that the mechanisms were the same. All of the evidence available for radiation-induced cellular damage indicated that it was an all-or-none (one-hit) phenomenon: Whether a DNA molecule within a cell was hit was a matter of chance, but if it was, it was damaged. An all-or-none mechanism precludes the existence of a threshold, a level of exposure below which no adverse effect will occur.

The use of radiation-induced mutagenesis as a model for chemical mutagenesis is the basis for the current controversy over whether chemical carcinogens have thresholds. If they behave like radiation, they have no threshold, and they would act as if they were chemical bullets producing mutations based on chance hits with DNA molecules. If they behave like chemicals, they have thresholds. In this case, they would follow the same anatomic, physiological, and biochemical pathways to reach and mutate DNA molecules as they do to reach any other site where they produce toxic effects.

In order to explain the similarities and the differences between chemical mutagens and X-rays, a brief description of X-rays and their physical and biologic properties will be helpful. X-rays are produced when a substance is bombarded by high-speed electrons. The X-ray machines that we see in doctors' and dentists' offices and in hospitals accomplish this by producing a high voltage in a vacuum tube to accelerate electrons across a short space to a metal target. Radioactive materials also produce gamma rays. X-rays and gamma rays differ from each other only in their source. X-rays and gamma rays are high-energy electromagnetic radiations that cause damage by ionizing the medium through which they pass. Ionization is the production of ions, atoms, or molecules that have an electrical charge, either positive or negative, as a result of the removal or addition of one or more electrons. In addition to X-rays and gamma rays, other radiations that can cause ionization include alpha particles, beta particles, and cosmic rays.

Radioactive chemicals (chemicals that emit ionizing radiations) complicate the comparison between the behavior of ionizing radiations and chemicals in their pathways through our bodies because they possess properties of both ionizing radiations and chemicals. They can do damage by virtue of their toxicity as well as by their radioactivity. The radiation damage also depends on their residence time in the body and the energy contained in the radiations they emit.

Radioactive chemicals are also called *radioactive isotopes*. Isotopes are atoms of a chemical element that have the same chemical properties but differ slightly from each other in their weights. Isotopes may or may not be radioactive. Those that are emit ionizing radiations of varying energies depending on the chemical identity of the isotope. The balance of this discussion will use only the terms *X-rays* or *ionizing radiation* because theories of mutation and cancer causation by ionizing radiation were based primarily on studies of the effects of X-rays and gamma rays, which behave the same as X-rays.

X-rays and chemicals differ markedly from each other in the way they enter and exit the body. X-rays are units of electromagnetic energy that pierce biologic systems in a straight line like bullets. They enter a body at any point on the surface, depending on the location of the radiation source. If radiations have sufficient energy to penetrate the skin, they may (1) pass through tissues and organs unchanged and exit the other side; (2) hit a molecule, lose some energy in the process, and continue on either to hit other molecules or to exit in a lowered energy state; or (3) lose all energy in a collision and be totally absorbed by the molecule. X-rays that pass through an organism without colliding with a biochemical structure lose no energy in the transit and do no damage.

From many studies, it is known that X-rays behave as extremely small units of energy and that the majority simply pass through tissue without interacting with it because, at the molecular level, our bodies are mostly composed of water molecules. Ionizing radiations must contact and transfer some of their energy to biological molecules if they are to modify them. The nature of damage to cells caused by ionizing radiations depends on how vital are the functions of the molecules they strike and modify. Ionization inside a living cell produces a sequence of reactions that result in damage to cellular components. If X-rays strike molecules critical to the health and integrity of a cell, the cell will sustain injury and may die if the damage cannot be repaired. If X-rays strike DNA molecules, mutations can occur and alter the genetic message.

Chemicals are units of matter and not units of energy; they do not strike the surface and travel through biologic systems in bullet-like fashion. Chemicals enter the body only if they can pass the barrier at the site of exposure—for example, lungs, skin, or digestive tract. Once transported across a barrier, chemicals follow well-defined anatomic, physiologic, and biochemical pathways through the body via veins and arteries to the organs and tissues. During its travels, chemicals meet many barriers, through which some can pass and some cannot.

In contrast to the situation with X-rays, the only randomness associated with a chemical pathway in a human organism occurs at the molecular level. The *individual* molecules of a chemical that cross a barrier do so in an apparently random manner governed by physical laws such as receptor binding, equilibrium, and osmosis. The total *number* and the *fate* of those that cross a barrier are not random but rather dictated by the biophysics and biochemistry of the organism. However, hits or collisions by X-rays with all sorts of extracellular and intracellular molecules and macromolecules, including DNA molecules, occur randomly and by chance.

The chemical bullet theory, patterned after the all-or-none theory for X-ray mutagenesis, postulates that one molecule of a chemical mutagen, like one unit of ionizing radiation, is capable of hitting a DNA molecule in a cell, thereby causing a mutation. Yet some chemicals are capable of altering DNA molecules by a chemical reaction rather than by physical strike. Such chemical reactions may involve oxidation (removal of electrons) of DNA, transfer of a piece of the chemical from itself to DNA, or transfer of a piece of DNA to the chemical. As with mutations produced by radiations, the fate of a mutated cell depends on the severity and location of the mutation.

The answer to the question of whether chemical carcinogens have thresholds is of tremendous importance for public health and the health and well-being of every one of us as individuals. Eventually the answer will be established through the study of the physiologic and biochemical mechanisms of the actions of chemicals.

CARCINOGENESIS

The word *carcinogenesis* is derived from the Greek words *gennan*, meaning "to produce," and *karkinos*, meaning "crab." The group of malignant diseases called *cancer* received their name from the Greek word "crab" because of their crablike attack on healthy tissue. Thus, carcinogenesis is the creation or production of cancer, and carcinogens are physical or chemical agents that cause cancer.

Cancer is one of the most dreaded of all the diseases that afflict the human race. Many other diseases are as or more life threatening than cancer, as destructive of the quality of life, and lead to more deaths in the population each year, but it is a rare person who will not breathe a sigh of relief when told that his or her illness is one of these other diseases and not cancer.

So it is with chemicals. The public has great concern for chemicals suspected of causing cancer while chemicals that may produce non-cancer diseases of lung, nerve, or kidneys are of less concern. As a result, suspected carcinogens are set apart from all other chemicals and made the subject of special laws and regulations.

Regulation of carcinogens is certainly appropriate. However, proper respect for all chemical products that one encounters in work or home environments, not just carcinogens, should be encouraged in order to prevent other severe and debilitating occupational diseases among workers, accidental poisonings of children in the home, and inappropriate use of prescription medicine.

What Is Cancer?

What is this disease called cancer that puts us into such panic, and what is known about the chemical causes? Cancers are malignant neoplasms.

Neoplasm, derived from the Greek meaning "new" and "formation," is the medical term used to designate any new or abnormal growth, such as a tumor that may be benign or malignant (i.e., one that does not spread throughout the body or one that does). Medical science knows a great deal more about the prevention and treatment of cancer than it does about its origin and preclinical development. Large groups of malignant diseases, possibly 100 or more, all with different etiologies and different prognoses are collected under the umbrella term *cancer*.

Malignancies are grouped together under the generic designation of cancer because they all have certain characteristics in common:

- *Cell proliferation.* Cancer cells grow much more rapidly than the normal cells from which they were derived.
- *Loss of differentiation.* Cancer cells lose some of the features typical of their normal parent cells.
- *Metastasis.* Cancer cells often invade and destroy adjacent tissues and usually spread to other more distant organs where they establish secondary cancers. One of the major differences between malignant and benign tumors is that the latter never metastasize.

There are four main types of cancers:

1. *Leukemias.* Cancers of certain white blood cells and the tissues from which they are derived.
2. *Lymphomas.* Cancers of tissues of the lymphatic system (e.g., Hodgkin's disease).
3. *Sarcomas.* Cancers of connective tissues, such as bone and cartilage.
4. *Carcinomas.* Cancers of epithelial tissues that form the inside and outside linings of our bodies (e.g., skin cancers). Carcinomas are the most common type of cancer.

Can cancer be inherited? Virtually all cancers have heritable and non-heritable forms, and a considerable body of evidence suggests that certain types of cancers, such as breast cancer, tend to run in families. However, it is difficult in many of these cases to rule out the influence of environmental factors as well.

The mechanism by which cancers cause death is not well understood. Cancers themselves are not thought to be the actual killers; they do not appear to release any toxic substances into the body. Solid tumors can, however, become large enough to physically block organs and tissues in the body, leading to impaired function. Cancers usually cause death by nourishing themselves at the expense of other tissues and organs, thereby causing the latter to become malnourished and subject to failure. The most common immediate causes of cancer deaths are infections, respiratory failure, hemorrhage, liver or kidney failure, and heart failure.

Causes of Cancer

What are the causes of cancer? For many years, researchers have recognized that there are geographic differences in the incidence of various types of cancers. They have also observed that when people migrate from one part of the world to another, within a few generations their offspring exhibit cancer incidences typical of their adopted land rather than those of the land of their origin. For example, in Japan, cancer of the stomach is much more common than cancer of the colon, whereas in the United States, the reverse is true. Japanese Americans have a stomach/colon cancer incidence ratio similar to that of their fellow Americans rather than to that of their parents.

The hypothesis that geographic differences in cancer incidence are due to environmental factors was first enunciated by Dr. John Higginson in the early 1950s. His hypothesis was based on an extensive comparative study of cancer among African black populations and black populations in other parts of the world. He concluded that 80 to 90 percent of all cancers were caused by environmental factors. With the growing concern about environmental contamination during the 1960s, it was only a matter of time before the term *environmental factors* became transformed into *environmental chemicals*. By the mid-1970s, statements to the effect that the majority of human cancers could be attributed to carcinogenic chemicals in the environment became commonplace.

Thomas H. Maugh II reported an interview in which Dr. Higginson explained that his conclusions have been misinterpreted "not among the majority of scientists with whom I have contact, but by the chemical carcinogen people and especially by the occupational people" (see "Research news: Cancer and the environment: Higginson speaks out," *Science* 205 (1979): 51). Dr. Higginson further stated in the interview that he viewed lifestyle as playing a much more important role than environmental chemicals in cancer causation. Dr. Higginson estimated that approximately 65 to 70 percent of all cancers are due to lifestyle, including the use of tobacco and alcohol. Of course, tobacco and alcohol contain chemicals, so separating lifestyle and chemicals as causative agents is a difficult task. Two other components of lifestyle are diet and behavior. An example of effect of diet is the association of colon cancer with lack of dietary fiber. An example of effect of behavior is the association of cancer of the uterine cervix with early onset of sexual activity leading to more frequent contact with the human papilloma virus, a known cause of cervical cancer.

Dr. Higginson said that, in addition to the cancers caused by diet and lifestyle, sunlight is responsible for approximately 10 percent; occupation for 2 to 6 percent; congenital factors for 2 percent; ionizing radiation for 1 percent; medical treatment for 1 percent; and the remaining 10 to 15 percent to unknown causes. Since his hypothesis was formulated, the causation of cancer by viruses has become accepted, accounting for much of the last category.

The hypothesis that a major fraction of human cancers had viral origins was formulated about 50 years ago, and many people hoped that eventually vaccines would be developed to immunize people against cancer. Several kinds of cancer support the viral theory; anticancer vaccines to prevent some types of cervical cancer have been approved in the United States, and other anticancer vaccines are in development. In addition, therapeutic vaccines are being developed to fight cancers by triggering the body's immune system to recognize cancer cells as foreign and to mount an attack against them. These latter vaccines treat cancers caused by a variety of mechanisms, and they are not the same as preventive vaccines.

The role of background radiation (cosmic rays, radon, etc.) in cancer causation may well be greater than is generally supposed. In fact, background radiation may be responsible for much of what is considered the background incidence of cancer from unknown causes. A visit to a science museum where devices that detect cosmic rays are exhibited reveals that we are constantly being bombarded with radiation of tremendously high energy. Theoretically, only one hit out of the billions that shoot through each of us is enough to initiate a cancer process. In any event, the background incidence of cancer is sufficient to make meaningful epidemiological studies of agents suspected of causing very small increases in cancer incidence, such as trace exposures to synthetic chemicals, extremely difficult, if not impossible.

Regardless of where the blame belongs, the people at greatest risk of developing cancer from exposure to chemical carcinogens are those who are exposed to the highest concentrations of these agents. The greatest and most varied exposures to chemicals of all kinds may occur in occupational settings.

Role of Mutation

The fact that many chemicals and physical agents that are carcinogenic are also mutagenic forms the basis of the theory that mutation sets the stage for cancer development. The scientific community has come to a general consensus that some, and perhaps even most, cancers do result from mutagenic events. Evidence also demonstrates that direct mutation is not the only mechanism. Some carcinogens, such as certain hormones, are difficult to fit into a simple mutagenic mechanism, and no perfect correlation exists between carcinogenicity of chemicals and their mutagenic activity in standard testing procedures.

The discovery of genes in human tumor cell chromosomes related to those in mouse tumors has stimulated an important avenue of cancer research. Previously, certain genes known as *oncogenes* had been shown to produce cancers in rodents. Oncogenes are the product of a mutating event that changes a proto-oncogene (usually a gene that codes for a protein that is important in the growth and programmed death of a cell) into an oncogene. The mutating event may be exposure to a virus, to an environmental chemical, or to radiation. The oncogene can protect a damaged cell from its natural death and lead to cancer. Further research has demonstrated the existence of tumor suppressor genes in the cancer process. Additional research indicates that activation of oncogenes and inactivation (or loss) of tumor suppressor genes are critical molecular steps in the transformation of a normal cell into a neoplastic (cancerous) one.

Incidence of Cancer

The incidence of cancer is dose related: The greater the dose of chemical carcinogen, virus, or radiation, the greater the number of individuals who will develop the cancer specific for that chemical. This is an extremely important point to remember when considering the carcinogenicity of chemicals. Many chemicals that occur as trace contaminants in the environment derive from some manufacturing or industrial process. Therefore, there must have been occupational exposures to these chemicals. Because of the very nature of the settings in which they occur and the quantities involved, these occupational exposures are tremendously greater than environmental exposures. If long experience with a chemical in occupational settings gives no evidence that the chemical is a carcinogen, it is extremely unlikely that exposure to trace quantities in the environment would cause cancer. If it did, the incidence level would be so low that it would not be separated out from the background incidence.

What is the incidence of cancer? According to the American Cancer Society (www.cancer.org), in 2010, approximately 570,000 people are expected to die from cancer in the United States—about 1,500 each day. The five-year survival rate from all types of cancer is 68 percent, which represents a huge increase since these types of statistics were first kept. However, some forms of cancers, such as skin cancer, have a much higher cure rate, whereas others, such as lung cancer, have a much lower cure rate. Cancer is the second leading cause of death after heart disease. Approximately one out of every four people in the United States will develop some form of cancer during his or her lifetime. An incidence of 25 percent is a compelling reason for every one of us to learn more about this group of diseases, their causes, and their prevention. The total incidence of cancer and the prevalence of the various types of cancer vary with geographic location, economic status, and cultural practices. In the United States, Caucasians have a lower incidence of cancer than the nonwhite population.

Categories and Characteristics of Carcinogens

The study of chemical carcinogenesis is complicated by the fact that they are varied in their interactions with living organisms. There is no systematic classification of chemical carcinogens. The designations used in the next paragraphs are for convenience in describing the various types of interactions; other authors may use different labels for the same interactions.

Chemical carcinogens may act as primary (direct) carcinogens, procarcinogens (indirect carcinogens), cocarcinogens, promoters, or secondary carcinogens. Primary carcinogens directly initiate a cancer process. They do not have to be converted to another chemical in the body before they are able to react chemically with DNA molecules and cause mutations; radiomimetic chemicals—chemicals that mimic ionizing radiations in their biological effects—are examples of primary carcinogens. Many of the anticancer drugs that are used in the treatment of cancer, alone or with radiation, are examples of primary carcinogens. It may seem strange that chemicals that can kill cancer cells can also cause cancer. There is good biological reason for this: Cells in the process of dividing (duplicating their chromosomes) are more susceptible to mutagenic agents than quiescent cells. Since cancer cells undergo division much more frequently than normal cells, they are much more subject to lethal mutations than normal cells, and so primary carcinogens are effective in killing the cancer cells.

Procarcinogens are not themselves carcinogens but may be converted metabolically to carcinogens. A host of natural and synthetic chemicals fall into this category, the most widely occurring of which is probably benzo[a]pyrene, a product of the combustion of organic matter. Foods that are burned or charcoal broiled are common sources of exposure by the general public to benzo[a]pyrene, which is metabolized to form an initiator. The major difference between primary carcinogens and procarcinogens is that the latter require biologic activation (i.e., metabolism) to be converted to a carcinogenic form. Some procarcinogens are converted to carcinogens only after undergoing several metabolic steps. Other procarcinogens have two or more pathways of metabolism open to them, one that does not produce a carcinogen and another that does. Vinyl chloride, which is known to be carcinogenic for humans, is an example of the latter type. Experimental data indicate that when vinyl chloride is present in the body in very small amounts, it is metabolized to vinyl alcohol by the same enzyme system that metabolizes ethyl alcohol. When the quantity of vinyl chloride exceeds the amount that the enzyme system can handle, another enzyme system comes into play, and the vinyl chloride is converted to vinyl epoxide, the carcinogenic form. People who drink alcoholic beverages appear to be more sensitive to the carcinogenic effects of vinyl chloride, perhaps due in part to the fact that ethyl alcohol is saturating the first enzyme system so that the vinyl chloride goes immediately to the second system, converting it to the carcinogenic form. Since the majority of chemical carcinogens are procarcinogens, we will refer to both primary and procarcinogens merely as carcinogens.

Cocarcinogens are neither carcinogens themselves, nor are they metabolically converted to carcinogens; rather, they enhance in some manner the carcinogenic activity of other chemicals that are carcinogens. Many natural and synthetic chemicals fall into this category. A cocarcinogen and a tumor promoter differ in that a cocarcinogen and a carcinogen must be present in the cell at the same time for cancer to occur, whereas a tumor promoter does not need to present concurrently with a carcinogen.

Several different mechanisms could explain the enhancement by cocarcinogens of the carcinogenic potency of other chemicals. Such mechanisms work by increasing the rate of conversion of procarcinogens to their carcinogenic forms, altering their biochemical pathways, increasing the rate of cell division, and interfering with repair mechanisms that undo damage done by carcinogens. For example, caffeine can act as a cocarcinogen by interfering with DNA repair mechanisms.

Promoters stimulate the growth of cells that have already been mutated by other agents. Some, such as croton oil, appear to play only this function and do not exhibit any other form of cocarcinogenic activity. Bile acids also seem to function as promoters only for colon cancer. Other chemicals, such a phenobarbital, DDT, and other chlorinated hydrocarbon compounds, may act as promoters in some situations and as cocarcinogens in others.

Secondary carcinogens set in motion a sequence of events that ultimately lead to some sort of tissue damage. The damage, which is not specific to the chemical, is considered to be the primary cause of the cancer, and the chemical is the secondary cause. Oxalic acid is an example of a secondary carcinogen. Chronic oral exposure to relatively low levels of oxalic acid, such as might occur with large daily intakes of foods containing high concentrations of the chemical, may result in bladder stones. The bladder stones in turn can irritate the lining of the urinary bladder, which can eventually lead to chronic damage and in turn to bladder cancer.

In an effort to explain differences between direct and indirect actions, researchers divided chemical carcinogens into two major categories, genotoxic and epigenetic chemicals. Chemicals that damage genetic material (DNA) and cause mutations directly, such as radiomimetic chemicals, were classified as genotoxic. Carcinogens that were not themselves capable of damaging genetic material were classified as epigenetic. The original distinction between genotoxic and epigenetic chemicals primarily distinguished between primary carcinogens and procarcinogens. However, this distinction was blurred by the discovery that procarcinogens could be converted metabolically to chemicals that can react chemically with DNA molecules. Many procarcinogens are now classed as genotoxic chemicals. The epigenetic category is reserved for chemicals that cause cancer by some mechanism other than mutation.

Some chemicals cause cancer in only one or a few species, in only one sex, or only in individuals with certain physical or genetic conditions. Many carcinogens are capable of causing cancer only by one or two routes of exposure and not by others. Sugar, for example, when injected under the skin of certain species of rodents, can cause a malignant tumor at the site of injection. Fortunately for people who have a sweet tooth, sugar has not been shown to be carcinogenic by ingestion in any species.

Chemical carcinogens are quite specific in the sites that they attack, usually producing malignancies in only one or a few sites. For example, aromatic amine compounds, such as the aniline dyes, cause urinary bladder cancer; mold toxins, such as aflatoxin, cause liver cancers; certain nitrosamines cause esophageal cancer; and thiourea causes thyroid cancer. Because of species difference in site of attack, when a chemical is labeled as a carcinogen, the public should ask what type of cancer and what increase in spontaneous incidence of that cancer it causes.

There is also evidence that deficiencies of certain chemicals can cause or predispose to cancer formation. In this category are chemicals that in some manner prevent or retard the aging process in cells. Cellular aging is recognized as an important factor in carcinogenesis. Thus, chemicals that retard cell aging in some manner may protect against cancer, and their absence may enhance aging and carcinogenesis. Vitamin C, vitamin E, beta-carotene, and selenium have been investigated for their anticancer properties. Obviously, it is meaningless to say that a chemical causes cancer by virtue of its absence, but the concept of deficiency plays a role in the carcinogenic process and underscores the tremendous complexity of chemical carcinogenesis.

The body's own hormones can influence the carcinogenic process. Several types of breast cancer are dependent on estrogen for the "feeding" of the cancer cells. For these cancers, specific drugs block the estrogen receptor, depriving the cells of a needed nutrient and thus slowing down cell growth and division. Similarly, prostate cancers are stimulated by the hormone testosterone and its active metabolite dihydrotestosterone (DHT), and there are drugs that block this hormone receptor.

Induction Periods

One of the difficulties in studying the causes of cancer and the carcinogenicity of chemicals is that cancers have relatively long induction times. The induction period, or latent period, is the time between the initiation of a cancer by the carcinogenic agent and the clinical appearance of premalignant or malignant symptomatology. The molecular events that occur during the induction period are not understood. Obviously, exposure to a carcinogen can set in motion a process that remains obscure for a period of time until clinical symptoms finally appear. A diagnosis of cancer may not even be made until long after cessation of exposure.

The length of the induction period for a given carcinogenic agent varies with dose: The greater the dose, the shorter the time between exposure and recognition of the disease. When exposure to a carcinogen is repeated day after day for many years, which is the usual case with occupational exposures, the latent period is obscured: Was the cancer initiated on Day 1 of exposure or after many years? It probably never occurs on Day 1, although a single overwhelming dose may be the cause in rare cases.

Some years ago, when one of us (PF) served on the Environment Board for a city, a man came before the board to demand that we stop adding fluoride to the water. Of course, our board had no authority to modify the municipal water supply, but we listened to his concerns. His primary evidence for the toxicity of fluoride as a carcinogen was the remarkable increase in the incidence of thyroid cancer in the year following the introduction of fluoride into the water supply of several U.S. cities in the early 1950s. His argument assumed that the induction period for fluoride as a carcinogen is extremely short. In reality, fluoride is not a carcinogen, but even if it were, the induction period would certainly be longer than the very short elapsed time between introduction of small amounts of fluoride into the water supply and the increase in thyroid tumors. Given the time frame of the increase in cancer, one could perhaps consider the detonation of the atomic bombs in the 1940s, both those tested in U.S. territories and those dropped in Japan, as the causative agent for increased cancer, since the induction time for a large dose of radiation to induce cancer is considerably shorter than that needed for chemicals, and the atomic bombs released radioisotopes that can be concentrated in the thyroid gland.

Data from experiments in which animals were exposed to carcinogens for different lengths of time, removed from exposure, and then followed for the balance of their lives show that a period of time (called a *preinduction period*) precedes the induction period. The total length of preinduction plus induction periods decreases with increasing exposure. An animal removed from exposure during the preinduction period will not develop cancer. The fact that preinduction periods exist is an extremely important point to consider when evaluating the potential for a brief, one-time exposure to a suspected carcinogen in humans. People who are accidentally exposed to low levels of chemicals suspected of being carcinogens, such as occurs frequently with polychlorinated biphenyls (PCBs) released during transformer accidents, are justified in assuming that they are not doomed by the episode. It has also been shown that longtime smokers who stop smoking revert to nonsmoking susceptibility to lung cancer over the course of years rather than remaining at the higher risk seen in smokers. The longer one smokes, the longer the time it takes for the lung tissue to return to baseline risk, since the insult to the lungs is cumulative over the years.

Information on lengths of induction periods for exposure to some occupational carcinogens were obtained from retrospective studies of workers in their postretirement years. However, the most reliable method for determining preinduction and induction periods is by controlled studies in laboratory animals in which exposure to known quantities of carcinogens is followed by long observation periods of no exposure.

Thresholds

No controversy exists regarding the existence of threshold doses for acute or chronic toxic effects, but a segment of the scientific community denies their existence for chemical carcinogens. The origin of this paradox lies in the use of ionizing radiation as the model for chemical carcinogenesis.

Current rejection of the threshold concept for carcinogens is rooted primarily in the unknown shape of the dose-response curve extrapolated from the lowest experimentally determined exposure to zero dose. If there is no threshold, then extension of the experimentally derived dose-response curve to zero effect would yield a curve that would go through the origin (zero

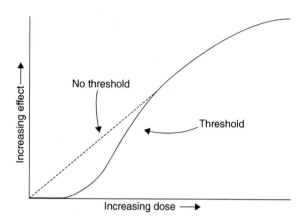

FIGURE 7-2 Dose-response curves extrapolated to zero for compounds with and without thresholds.

dose). If there is a threshold, then the extended line would meet the axis at some point greater than zero dose (see Figure 7-2).

In 1971, the National Center for Toxicological Research (NCTR) was created to provide federal regulatory agencies such as the Food and Drug Administration (FDA) and EPA with the scientific data they required for the performance of their regulatory duties. One of the first missions of NCTR was to shed some light on the question of thresholds for carcinogens. To this end, in 1972, the center designed an animal carcinogenicity feeding study of hitherto unimagined size. Its purpose was to describe accurately an ED_{001} for a known carcinogen whose effects had been extensively investigated and were relatively well understood. *ED* means "effective dose" (in this case the effect is cancer), and 001 means an effect for 0.1 percent (one per 1,000) of the animals in the study.

Such a study, it was hoped, would provide experimentally derived points lower on the dose-response curve than had ever been obtained before and, in doing so, shed some light on the threshold question. The physical resources of NCTR, however, could not accommodate the number of animals required by an ED_{001} study, so the plans were scaled down to an ED_{01} study (effective dose for 1 percent of the animals).

The carcinogen selected was 2-acetylaminofluorene (AAF), a chemical for which a great deal of carcinogenicity data was already available. AAF fit the criteria established by NCTR for the test chemical better than any other known carcinogen. The study required 18 months for the planning stage, 9 months for production of the more than 24,000 mice employed in the study, another 3 to 4 years to conduct the study and evaluate the data, and a cost of somewhere between $6 and $7 million. If this study were performed today, it would cost a great deal more. (The animals alone would cost about $1 million.) The results of this heroic effort were published in a special 1980 issue of the *Journal of Environmental Pathology and Toxicology* (vol. 3, pp. 1–250).

Evaluation of the massive quantities of data produced by the ED_{01} study required years of effort following the completion of the experimental phase. The study produced much information extremely important to the design and conduct of future carcinogenicity studies, and it contributed immensely to the development of appropriate models and formulae for carcinogenic risk assessment. However, the question of whether a threshold exists for the carcinogenic effects of AAF was not answered. Thus, the question of whether thresholds exist for carcinogens still remains, and may always remain, unanswered.

Federal regulatory agencies adopted the no-threshold theory of carcinogenesis many years ago when faced with the requirement to make regulatory decisions about chemicals classified as carcinogens, whether proven or suspected. The policy was based on the assumption that since radiation "strikes" in biological materials were stochastic (random) events, chemical carcinogen "strikes" were also stochastic. In stating its policy, EPA wrote: "stochastic effects are those for which the probability of an effect occurring, rather than its severity, is regarded as a function of the dose without threshold" (*Federal Register* 44, no. 52 (March 15, 1979): 15,975). For such effects, which may be regarded as all-or-none phenomena, thresholds of no effect levels cannot be established because even extremely small doses must be assumed to elicit an increase in the incidence of the response. Carcinogens, mutagens, and in some cases teratogens elicit stochastic effects. Consequently, safe levels—that is, levels that will produce no adverse effects—cannot be established for carcinogens and mutagens.

As an example of how this controversy continues, in 2000, EPA evaluated data on the carcinogenicity of dioxin using the low-dose linear extrapolation method and concluded that dioxin is a carcinogen that has a threshold. A reanalysis of the data in 2003 by Mackie and others showed no threshold ("No evidence of dioxin cancer threshold," *Environmental Health Perspectives*, 111 (July 2003)). Thus, the threshold question remains under debate for dioxin and many other carcinogens.

Practical Thresholds

Regardless of the debate over the existence of thresholds for chemical carcinogens, the public should be aware that there are practical thresholds for all carcinogens. There is no disagreement in the scientific community that the incidence of carcinogenic effects and the length of their induction periods are dose related. The greater the exposure to a carcinogen, the greater the number of people in whom it will cause cancer. Therefore, if exposure to a carcinogen is sufficiently small as to reduce its cancer incidence to one in a trillion, that level of exposure would be of no practical significance for humans because there are not that many people in the world. The chances of getting cancer from that carcinogen would be infinitesimally small. Practically speaking, this would also be true if the incidence was still small but somewhat larger. We address these issues for cancer as well as other toxicities in Chapter 11 when risk benefit is discussed.

Although the exposure is directly related to dose, the induction period for is inversely related. The larger the dose, the shorter the induction period, and the smaller the dose, the longer the induction period. If the level of exposure to a chemical carcinogen is sufficiently low as to increase its induction period to 200 years, of what practical significance would that be for the human population? The human race has not yet achieved an average life span of 100 years, much less 200.

THE REAL WORLD

Benzo[a]pyrene

All humans are exposed throughout their lifetimes to countless numbers of chemical carcinogens, many of which are naturally occurring, yet not all humans, or even most, develop cancer. The quantities of carcinogens to which we are exposed are well in excess of practical thresholds. To illustrate, benzo[a]pyrene is a naturally occurring and relatively potent carcinogen that is virtually omnipresent in our environment as a product of the cooking or burning of any organic material. It has been determined that there are 50 µg of benzo[a]pyrene in about 2 pounds of charcoal-broiled steak. A generous portion of steak would weigh about one-fifth of a kg (7 ounces [oz]) and thus would contain about 10 µg benzo[a]pyrene. Ten µg is a very small quantity (there are over 28 million µg in 1 oz). However, the number of molecules contained in 10 µg is huge. (There are 24,000,000,000,000,000 molecules in 10 µg of benzo[a]pyrene.) That is, about 24 quadrillion molecules of benzo[a] pyrene are contained in an average portion of charcoal-broiled steak (See Appendix A for how to calculate the number of molecules)

We do not eat charcoal-broiled steak every day, but considering that benzo[a]pyrene is so widespread in our environment, it is reasonable to assume that each of us ingests at least billions of molecules of benzo[a] pyrene every day (a billion is only one millionth of a quadrillion). The most common site for cancer caused by ingestion of benzo[a]pyrene is the stomach, yet in the United States, stomach cancer is one of the less common cancers, accounting for only approximately 0.01 percent of all malignancies (<25,000/year). However, benzo[a]pyrene in cigarette smoke may possibly contribute to the increased incidence of lung cancer found among cigarette smokers.

If the simple one-hit or chemical bullet mechanism for chemical carcinogenesis were a reality, and if there were no thresholds for chemical carcinogens, what would be the practical consequences? Benzo[a]pyrene is only one of a host of natural and synthetic chemical carcinogens to which we are regularly exposed. Even minuscule exposures represent billions and billions of molecules. If each molecule of a chemical carcinogen behaved as a unit of ionizing radiation, it is difficult to understand how anyone could escape multiple cancers, much less one cancer.

Cancerophobia

The fear of cancer is one of the greatest fears that people have about exposure to synthetic chemicals, especially in pesticide residues and food additives. It is important for people to remember that someone synthesized these chemicals and someone worked with them, usually long before they became available for contact by the general public. As a rule, people in those occupations have been exposed to much higher concentrations of these chemicals and for longer periods of time than the average person. There is no chemical that can cause cancer whose carcinogenic properties remain hidden for very long after coming into general use in industry. The chance of a chemical causing neurological or developmental disorders is much harder to detect than the causation of cancer.

Even though prevention of all occupational disease, including cancer, is the major goal of the occupational health professions, the occupational setting probably will always provide the greatest source of human exposure to potentially dangerous synthetic chemicals. If a chemical does not cause cancer in exposed occupational groups, it is highly unlikely that it will cause cancer in the general population when it is administered by a similar route. However, we know that most occupational workers do not swallow the medicines they manufacture or inhale smoke when making cigarettes. The exposure routes for these types of chemicals may not be determined in the user population by monitoring occupational workers, hence the length of time needed to establish the association between tobacco use and lung cancer. If tobacco workers had shown lung cancer from handling tobacco, we would have seen this relationship much earlier; however, it is not the tobacco itself that is causative, but the chemicals released during burning or chewing that are the carcinogens.

When the route of exposure is the same in workers as in the general population, occupational workers really become the canaries in the mine for the rest of us. And this is literally true in the case of coal miners who suffer not from occupationally induced cancer but from "black lung disease" (coal worker's pneumoconiosis), an inflammation and fibrosis of the lungs caused by inhaling the carbon and silica dust in coal. The relationship between the disease and the cause has been known since humans first began to mine coal, and the temporal relationship between exposure and time to onset of this disease and its severity are understood by all those who live in coal-mining communities.

CHAPTER 8

DEVELOPMENTAL AND REPRODUCTIVE TOXICITY

The influence of chemicals on reproductive function in both men and women has been the subject of extensive research for many years. Much of this research has involved the development of methods to interfere with reproductive function. However, the protection of reproductive functions and of the developing fetus are also important research efforts.

In past years, before modern methods of contraception were available and before legal abortions were available in the event of failure of these methods, self-induced abortion was a social and medical problem of far greater magnitude than it is today. Some women resorted to naturally occurring abortifacients. Substances such as tansy oil, turpentine, nutmeg, ergot, pennyroyal, and oil of savin (a toxic oil extracted from a juniper species and used as far back as ancient Rome) have been used since to induce abortion, often with fatal results to the mother. The abortifacient actions of these substances are unreliable, unpredictable, and unsafe when ingested or applied vaginally in quantities sufficient to terminate pregnancy. Unsuccessful use of these chemicals frequently led to severe damage to the fetus.

On a botanical tour through any tropical forest, the majority of useful plants pointed out by indigenous guides are those affecting the reproductive system: to act as an aphrodisiac, to interfere with pregnancy through prevention or early termination, or to stop postpartum bleeding. Having taken some of these tours, we estimate that fully 90 percent of the plants mentioned fall into one of these categories, with the remainder used to treat fever, to counteract poisons from other plants and animals, and to heal wounds.

As toxicologic sophistication has increased through the decades, so has recognition that exposure to some chemicals can produce undesired effects more subtle than previously imagined. Unfortunately, alterations in the reproductive process may be counted among such subtle effects. A large number of points in the human reproductive process are vulnerable to interference by physical or chemical agents. This process encompasses the

The Dose Makes the Poison: A Plain Language Guide to Toxicology, Third Edition.
By Patricia Frank and M. Alice Ottoboni
© 2011 Patricia Frank and M. Alice Ottoboni. Published 2011 by John Wiley & Sons, Inc.

153

development of the male and female reproductive systems, the maturation and fulfillment of their functions, and the development and viability of new individuals created by the process. Reproductive toxicology deals with adverse effects caused by chemicals at any point in the process. Because of the specific nature and complexity of its purview, reproductive toxicology is a well-populated subspecialty within the field of toxicology.

MALE AND FEMALE REPRODUCTIVE SYSTEMS

Prepuberty Systems

Under normal conditions, the sex of a new individual is determined by a pair of chromosomes known as the X and Y, or sex, chromosomes. Female germ cells (ova, or eggs) contain one X chromosome. Male germ cells (sperm) are of two kinds; they contain either one X or one Y chromosome. When a sperm and egg unite, a female will result if the sperm contributes an X chromosome to pair with the X chromosome in the egg; a male will result if the sperm contributes a Y chromosome. That is, human females are XX and males are XY.

Early in the life of a developing embryo, a group of cells is set aside to become the reproductive system of the adult. During embryonic and fetal development, the gonads (ovaries or testes) and accessory sex organs (e.g., clitoris and prostate) are formed. In a female, the reproductive system remains within the body, and a canal (vagina) to the exterior is formed. In a male, the system is externalized, and the testes descend into the scrotal sac. The migration of the testes into the scrotum is an important step in male sexual development. The testes must be maintained at a temperature several degrees cooler than that in the abdominal cavity for the delicate process of sperm production to occur. A male whose testicles do not descended until puberty, although sexually normal in all other respects, usually will be sterile.

At birth, the male gonads contain primitive germ cells (spermatogonia). Through a complex process of division and differentiation, spermatogonia produce many billions of sperm, beginning at puberty and continuing throughout the reproductive life of the man. The female gonads contain a small number of potential eggs, which is determined before birth. The primitive germ cells in the ovaries (oogonia) undergo the first step of development during fetal life to form primitive eggs (oocytes). Each month, beginning at puberty, one oocyte usually will develop into a mature egg and be released in a process called ovulation. On occasion, two or more eggs will be released at the same time. Without the use of drugs to promote ovulation in the treatment of infertility, the release of more than two eggs during one ovulatory period is rare.

There has been relatively little study of the adverse effects of environmental chemicals on the development of the male and female reproductive systems from conception to birth and on the maturation and competence of their functions from birth to puberty. Research in this area has revolved primarily around effects of the internal environment, such as hormonal imbalances, disease states, and physical agents, such as radiation.

The impact of temperature on sperm production has long been known. Any factor that increases testicular temperature in mature males, such as tight jeans that hold the testicles too close to the body and perhaps even prolonged exposure to excessive environmental temperatures such as steam baths, can cause reduced fertility or even sterility. Male sterility caused by the disease mumps is considered to be due primarily to elevated scrotal temperatures; fortunately, when prepubescent boys get mumps, sterility is much less common than for adults.

The discovery years ago that some cases of vaginal cancer in prepubescent girls were due to their exposure in utero to the synthetic hormone diethylstilbestrol (DES) given to their mothers during pregnancy illustrates the vulnerable steps in the reproductive process prior to puberty. The DES findings, together with those of the antinausea drug thalidomide discussed later in this chapter, led the Food and Drug Administration (FDA) and worldwide bodies to impose stronger regulatory requirements on pharmaceutical companies to study the effects of drugs on the reproductive system. The methods used for these studies are discussed in Chapter 5.

Adult Systems

Puberty, accompanied by production of sperm in the male and the onset of menses and egg production in the female, marks the beginning of human reproductive life. Until recent years, the preponderance of research on the human reproductive process has been directed toward the female with the goal of developing and perfecting more effective methods of birth control. Efforts to develop male contraceptive drugs have been ongoing for many years with little success for an acceptable method. The search for contraceptives has added greatly to the knowledge of female and male reproductive biology.

After puberty, many more steps in the reproductive process can be impaired by physical or chemical agents. The male must be able to produce good-quality sperm of proper size, shape, and motility. He must produce them in sufficient quantity to be fertile—an average of many millions in each ejaculation. He must be able to produce adequate transport fluids for the sperm and be capable of delivering them into the female. The male reproductive process terminates when the sperm leaves the body.

The adult female must be able to produce a good-quality egg and release it from the ovary. The egg must be swept into a fallopian tube, the viaduct that delivers the egg to the uterus, where fertilization usually takes place. The egg must be able to permit penetration of a sperm cell. If the egg is not fertilized, it passes out of the body, and the monthly ovulatory cycle begins anew.

If the egg is fertilized, it must be maintained as it travels down the fallopian tube to the uterus. The first few cell divisions occur during this time. Once the fertilized egg reaches the uterus, approximately a week after conception, it must be able to implant itself in the uterine wall and form a connection with the maternal blood supply so that it can continue to be nourished and grow. The uterine wall must be receptive to the little group of cells destined to become a new individual. The female role in the reproductive process continues through birth and weaning of the infant.

Considerable data in the scientific literature demonstrate the sensitivity of oocytes to destruction by X-rays. Benzo[a]pyrene, a chemical widely distributed in nature, is a product of burning organic matter and is capable of destroying oocytes. The effects of chemicals on male and female germ cells was of little scientific interest before the discovery in 1977, and its accompanying national publicity, that the pesticide dibromochloropropane (DBCP) was capable of making men sterile.

THE DEVELOPING INDIVIDUAL

At the time a fertilized egg reaches the uterus and implants in the uterine wall, it is a little ball composed of cells that are actively dividing. When this ball of cells implants, it officially becomes an embryo and is so called until the end of the eighth week of life. From the beginning of the ninth week until birth it is called a fetus. After implantation, the embryo forms a sac (amniotic sac) around itself that becomes filled with the fluid in which the fetus will "float" until birth. A communication system consisting of the umbilical cord and the placenta becomes functional around the third week of life. Prior to that time the embryo relies on the diffusion of products from the uterus for nourishment.

The placenta forms an intimate contact with the wall of the uterus, and the area of contact between the two is called the placental barrier. Blood cells do not cross the barrier; maternal blood and fetal blood do not intermingle. Nutrients pass through the barrier from the mother's blood to fetal blood, and waste products pass in the opposite direction. Chemicals vary in their ability to cross the placental barrier. Even some biochemicals, such as certain of the mother's hormones, do not cross to the fetus, at least not in sufficient quantities to interfere with fetal development. However, it appears that many pharmaceutical products do cross the placental barrier. New research is demonstrating that the barrier is not as selective as was once thought.

The human embryo, like all other vertebrate embryos, develops from head to tail. The primitive brain, spinal cord, and heart are among the first identifiable structures to appear. By 8 weeks, the head is formed, with eyes, nose, and mouth mapped out, and the upper torso has primitive arms, hands, and fingers. The lower torso is still forming, and only little buds indicate where legs eventually will develop. By 12 weeks, the end of the first trimester, all organ systems have formed and begun to function. Fetal growth and refinement of physical features and organ functions continue until birth.

During the nine months of uterine life, there are innumerable developmental steps that are subject to interference or disruption. Congenital malformations (anomalies, abnormalities), more commonly referred to as birth defects, may result from such an interruption. The normal incidence of congenital malformations from all causes in the United States is estimated to be about 2 in 100 live births, and probably about 10 in 100 if mental deficiency is included. Approximately 20 to 25 percent of all pregnancies do not reach full term but result in spontaneous abortion, primarily due to fetal abnormalities. That so many newborn infants are born whole and healthy is amazing.

A host of factors may cause birth defects: disease or malnutrition of the mother, genetic abnormalities, or exposure of the embryo or fetus to physical or chemical agents. Some congenital malformations are so severe that they result in fetuses that are grossly deformed and nonviable, often resulting in spontaneous abortion. Other malformations that are compatible with extra-uterine life run the gamut from mild to severe. A defect may be readily detectable at birth or may not be manifest until several or many years later. When an infant is born with a birth defect, it is often very difficult or impossible to determine what factor or factors were responsible.

One of the causes of human congenital abnormalities is disease in the mother, particularly viral diseases such as rubella (German measles) during the first three months of pregnancy. Another frequent cause is heredity. Inherited abnormalities are traits that exist in the gene pool, perhaps from mutations that occurred very much earlier in the family line and find expression when individuals carrying the traits come together as parents. Nutrition plays an important role in the success of a pregnancy: Starvation, malnutrition, or deficiencies (or excesses) of certain vitamins or amino acids can interfere with normal fetal development. Maternal age at the time of conception is also related to the incidence of congenital anomalies; the chances that a baby is malformed begin increasing at about age 30 and take a dramatic upswing after the age of 40.

The process whereby some foreign agent, physical or chemical, produces an abnormality in a developing organism during uterine life is called teratogenesis, a word that comes from the Greek words *gennan*, meaning "to produce," and *terata*, meaning "monster." Thus, teratogens are agents that give rise to malformed or otherwise abnormal fetuses. Some chemicals and certain physical agents, such as natural and human-made radiation, can behave like teratogens. Chemical teratogens may or may not be mutagens, and chemical mutagens are not necessarily teratogens. X-rays, in particular, are both potent mutagens and teratogens. The use of X-rays during pregnancy, once a popular obstetrical practice to determine the position of the fetus, number of infants, and so forth, is now reserved for very special circumstances. Fortunately, it appears that ultrasound, which has taken the place of external radiation and is used to obtain a fetal sonogram, does not have a teratogenic effect.

In order for a congenital abnormality to be produced by a teratogenic agent, the agent must physically contact the developing organism. In the case of ionizing radiations, the embryo or fetus must be in the path of the radiations as they pass through the mother. In the case of viruses or chemicals, they must be capable either of crossing the placental barrier or of producing teratogenic substances (toxins or metabolites) that can cross.

A teratogen must not only be in contact with the developing organism, it also must reach it during the critical phase in the gestation period when organ systems are in the process of formation. The dividing egg is considered to be protected fairly well against the teratogenic action of chemicals during the week to 10 days from conception to implantation because of the lack of a direct communication system with the mother. The critical stage of organ development for the human is the first trimester of pregnancy. The kind of abnormality produced depends on what organ system is undergoing the most rapid development at the time of exposure to the teratogen. For example, exposure to high doses of vitamin A on the eighth day of gestation in rats results in skeletal malformations, whereas the same dose on the twelfth day results in cleft palate.

Certain steroid hormones, vitamin A, vitamin D, some drugs (especially anticancer therapeutics), and, of course, thalidomide are all known teratogens. A number of other chemicals are suspected of being teratogenic; many of these are industrial chemicals or drugs, about which physicians can provide information. For women who require prescription medicine to treat ongoing conditions, consultation with their doctors before becoming pregnant is a good idea. Additionally, sufficient levels of folic acid (vitamin B9) started before conception can significantly reduce the incidence of spina bifida, a birth defect found in approximately 1 to 2 live births per 1,000 pregnancies. Although grain products in the United States have been fortified with folic acid since 1998, the amount required is usually achieved by a vitamin supplement. By the way, folic acid at the required dose is a prescription product—one over-the-counter folic acid tablet does not contain nearly enough active compound to meet the supplement need. The prescription is, however, quite inexpensive.

The wisest course for any woman who is pregnant, particularly during the first three months, is to observe all rules of good nutrition and avoid smoking, alcohol consumption, and exposure to as many chemicals and medicines (including over-the-counter drugs) as possible. If work or hobbies involve the use of chemicals, a woman should seek her obstetrician's advice concerning such exposures. One of the difficulties with such advice is that a woman often does not know she is pregnant until after some important organs have already been mapped out in the embryo. This forms the basis for the dilemma involving women's rights and fetal protection. The question of whether women of childbearing age should be excluded from certain jobs involving exposure to industrial chemicals in order to prevent possible harm to an unrecognized embryo was presented to the Supreme Court for adjudication. In 1991, the Supreme Court ruled that "employers may not bar women of childbearing age from certain jobs because of potential risk to their fetuses" (see *United Auto Workers v. Johnson Controls Inc.*, U.S. Sup. Ct, No. 89–1215, petition filed January 29, 1990). The protection of the fetus in this case was adjudicated less important than sexual discrimination. In this particular case, the chemical in question was lead, which was in higher concentrations than deemed safe, but this decision applies more broadly; it is thus important for women to assess their own risk.

Chemicals also can produce fetal malformations by mechanisms other than direct effects on the fetus. They may produce mutations in parent germ cells before conception, which in turn result in a birth defect. They may interfere with development of the placenta and thus prevent adequate transfer of nutrients from mother to fetus. They may prevent adequate oxygen transfer to the fetus. They may interfere in the growth of organs or organ systems, or the refinement of their structures or functions, resulting in an anatomic, physiologic, or biochemical defect. They may exert a toxic action on the fetus and thereby cause degenerative changes in one or more anatomic, physiologic, or biochemical systems. Cocaine use during pregnancy, for example, may lead to premature birth, enhanced risk of miscarriage, and neurological defects, such as delayed learning and inability to pay attention. These effects usually are produced by cocaine use during the second and third trimester after organogenesis is complete; children look normal but behavioral difficulties arise later. For a fuller discussion of the effects of cocaine as well as other drugs see International Birth Defects Information Systems (IBIS) (www.ibis-birthdefects.org).

The fetotoxic chemical that probably is responsible for more fetal abnormalities than all other chemicals combined is alcohol, even when taken in the small amounts described as social drinking. Alcohol also may produce acute or chronic intoxication in the mother, causing a generalized debility in her and, secondarily, some adverse effect on the developing fetus. The latter situation is probably rare in humans, except in cases of excessive use of alcohol, illicit drugs, and tobacco, but it can be readily achieved in laboratory animals.

A congenital anomaly resulting from the action of a mutagenic agent on a parent germ cell or on an embryonic cell may result in a changed message for some relatively minor structure or function that may in turn result in some major defect. Depending on the nature of the change, the mutation may:

1. Never be expressed during the lifetime of the individual.
2. Show itself in some physical deformity or disease state compatible with survival to adulthood.
3. Result in a deformity or disease that is too great to permit long-term survival so that the individual dies at birth, in infancy, or in early childhood.
4. Cause death of the fetus in utero.

For a congenital malformation to be inheritable, the message for the defect must be carried in the germ cells of the malformed individual. Thus, the mutation must have occurred at some time prior to the segregation of the cells destined to become the individual's reproductive cells. The mutation could have occurred sometime in the distant past, with the defect carried in the germinal line of the ancestors down to the sperm or egg of a parent, or in the immediate past by action of a mutagen on a reproductive cell of a parent, or early in the life of the fertilized egg, before differentiation of the cells destined to become germ cells. Malformations resulting from mutations of somatic cells of the fetus will not be inherited by the progeny. A mutation in a germ cell will become part of the human gene pool only if the individual carrying it is able to survive to adulthood and reproduce.

Compounds also can act on the developing fetus, newborn, infant, and child to produce a variety of developmental effects that are not due to mutagens or teratogens. For example, exposure to lead any time from before birth through childhood can result in delayed development and permanent damage to the central nervous system. It has been proposed by researchers at Johns Hopkins University and the University of Michigan, among others that mental decline with aging can be attributed partially to childhood exposure to lead. It is estimated that approximately 20 percent of U.S. children have blood lead concentrations above acceptable levels. During 2008 and 2009, many cases of lead appeared in toys (usually in the paint) imported from China (doll shoes, various children's jewelry, blocks, wooden railway toys). Because children put items in their mouths and because they are most at risk for the effects of lead toxicity, it is imperative that local, state, and federal agencies continue to enforce the lead limits in consumer products. However, perhaps a larger tragedy concerning lead is that many children in China are being exposed to large amounts of this neurotoxin from air and water expelled from factories and from spilling or dumping slag from smelting plants. Many of these children have been hospitalized with signs of severe lead poisoning. The lack of regulations on the disposal of toxic materials has led to unacceptable pollution levels of lead in the air and water in many parts of China. This is of concern not only for the long-term health of Chinese workers but for the whole world since we know that no pollution is a local problem—it is all global.

CHAPTER 9

CASE STUDIES IN TOXICOLOGY

It is unfortunate that much knowledge of the chronic effects of chemicals has come to us from the illnesses of people who worked with them. These people served as guinea pigs in the laboratory of human experience. Today we would hope that society is more enlightened and that workers would not be permitted to be exposed to chemicals before the potential consequences were known. The hope is fulfilled in part because there are now improved methods for studying the chronic effects of chemicals, and the public and government are aware that more must be known about these effects. As a result, laws have been enacted that require the conduct of chronic toxicity tests before large populations of people are knowingly exposed to many potentially harmful chemicals. However, for food supplements, there are still few or no requirements for premarket testing, and for medicines, clinical trials usually consist of a few hundred to a few thousand patients, which may not be enough to discover rare side effects.

However, we would be deceiving ourselves if we believed that scientific expertise, awareness of the problem, and laws requiring chronic toxicity testing can be as protective of people as we would want them to be. The problem is not lack of desire or motivation to protect the population but rather the nature of toxicological science itself. No matter how much effort is made to study of the chronic adverse effects of chemicals, and no matter how many studies yield negative results, one can never be sure that some subtle effect has yet to be discovered.

The importance of human data in evaluating the potential for adverse effects from exposure to chemicals, both acute and chronic, cannot be overstated. Whenever human data are available, they must take precedence over animal data. After all, what animal model approximates human physiology and biochemistry more closely than the human itself? Unfortunately, data from human experience may be ignored or rejected by nonexperts in

The Dose Makes the Poison: A Plain Language Guide to Toxicology, Third Edition.
By Patricia Frank and M. Alice Ottoboni

toxicology, particularly when the data indicate that humans are less sensitive than laboratory animals or when the matter becomes a political issue rather than a scientific one. The most valuable human data come from tragic accidents or from mass exposures of people, either deliberately or inadvertently. We discuss just a few examples of the many such episodes.

ENVIRONMENTAL CONTAMINATIONS

Dioxins: Seveso, Italy

The term *dioxin* has become a prominent entry in our lexicon of environmental pollutants. This is because dioxins, as contaminants in certain major industrial chemical products, are among the most toxic chemicals that have ever been synthesized, albeit inadvertently, by humans. They were responsible for outbreaks of chick edema disease in the late 1950s that decimated large numbers of broiler flocks fed contaminated feed. They were responsible for the deaths in the early 1970s of many horses in a Times Beach, Missouri, horse farm where dioxin-contaminated oil was used to settle dust. They were the contaminants in the infamous herbicide Agent Orange blamed for numerous medical problems suffered by Vietnam veterans. During the Vietnam War, over 19 million gallons of Agent Orange were released over South Vietnam in a 10-year period. Even today some children in Da Nang, Vietnam, where much of the Agent Orange, was stored have body levels 50 times higher than the World Health Organization (WHO) standards. These children were not born until long after the war, a fact that demonstrates the persistence of dioxin in the environment.

Dioxins, or chlorodibenzodioxins, as they are more properly called, are a class of compounds formed as by-product contaminants in a number of chemical reactions, including combustion. They are formed in trace quantities during the manufacture of some chlorinated aromatic organic compounds, such as 2,4,5-T and 2,3,7,8-T (components of the herbicide Agent Orange), trichlorophenol, or pentachlorophenol. Chlorodibenzodioxins are chlorine derivatives of the basic dibenzodioxin (DD) molecule, which contains 12 carbon and 2 oxygen atoms.

The 12 carbons are arranged in two 6-carbon rings that are connected to each other by two oxygen bridges that join two adjacent carbons on one ring to two adjacent carbons on the other ring (see Figure 9-1). The DD structure can accept from 1 to 8 chlorine atoms—one chlorine for each of the carbons not tied up with oxygen. There are dozens of possible combinations of chlorine with DD. For example, there are two possible monochloro DDs, 10 possible dichloro DDs, 14 possible trichloro DDs, and so forth. However,

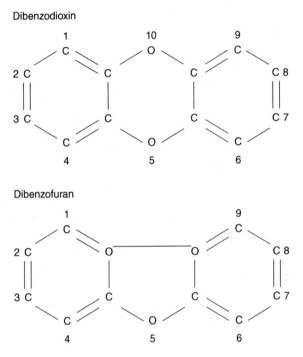

FIGURE 9-1 Dibenzodioxin and dibenzofuran structures.

there is only 1 octachloro DD because there is only one way to attach 8 chlorines to the 8 carbons.

Environmental concerns about DDs center on their generation by a variety of human activities. The theory that chlorinated DDs and dibenzofurans (DFs) could be produced from chlorinated aromatic hydrocarbons by the action of heat or ultraviolet light was suggested years ago by Donald G. Crosby of the Department of Environmental Toxicology at the University of California, Davis. The subsequent discovery that chlorinated DDs and DFs are produced in trace amounts during the combustion of solid waste materials demonstrated that Dr. Crosby's hypothesis was correct.

The importance of these findings to the management of land and air quality is great because the burning of organic materials results in trace quantities of dioxins in the ambient air in populated areas. This background level of dioxin contamination, combined with the fact that rapidly dwindling sanitary landfill reserves mandate the development of alternate methods of solid waste disposal, has created a dilemma for local governments.

Recycling of dry materials, such as metal, plastic, glass, and paper, is a serious endeavor in many communities. However, even if every household participated, recycling would eliminate only a small fraction of the garbage produced. One of the most cost-effective and efficient methods of waste disposal is conversion of waste to energy in power plants. In fact, such conversion plants are significant sources of revenue for the communities that have built them. However, public awareness that the process has the potential for producing chlorodioxins, even though it may add negligible quantities to background levels, has led to rejection of these incinerators by some communities. Public fears of chlorodioxins far outweigh public concerns for the looming garbage crisis; whether this is wise or not is an open question.

Toxicity studies of chlorodioxins indicate that the intensity of their adverse effects increases from the monochloro derivatives up to the tetra-chloro and then decreases from tetrachloro up to the octachloro compound. One tetrachloro appears to occur more commonly than the others: 2,3,7,8-tetrachlorodibenzodioxin (TCDD for short). TCDD appears to be the most toxic of the chloro DDs. In studies of the acute toxicity of TCDD, the guinea pig is the most sensitive animal that has been discovered thus far. The oral LD_{50} for the guinea pig is between 0.6 and $4 \mu g/kg$. The chicken appears to be the next most sensitive, with an oral LD_{50} between 25 and $50 \mu g/kg$. The LD_{50} values for all other animals, rats, mice, dogs, and so on, are between 100 and $200 \mu g/kg$.

Chronic toxicity studies in monkeys show that 50 ppb TCDD in the diet causes death after 2 months, 5 ppb TCDD in about 6 months, and 0.5 ppb in about 11 months. TCDD is a potent teratogen in rats and mice, but it does not appear to be a teratogen in monkeys. It is too toxic in guinea pigs to conduct successful teratology studies in them. The toxicity of TCDD shows wide variation among species in both potency and in the organ systems involved. A major chronic toxic effect of TCDD in humans is chloracne, a skin disease resembling the acne common to adolescence but much more persistent and somewhat more severe. There may be other effects in humans, such as liver damage or increased risk of certain cancers; however, the data are insufficient to determine whether these are associated with TCDD or with the chemicals that have TCDD as a contaminant. Investigations of an association between TCDD exposure and adverse effects on the human reproductive process have been inconclusive.

The carcinogenicity of TCDD is difficult to assess. In animal studies, it appears to increase the incidence of some tumors and decrease the incidence of others. The fact that it influences the incidence of a variety of cancers supports the thesis that it is not a carcinogen but rather a cocarcinogen or promoter.

How might the data on the toxicity of TCDD obtained from animal experiments translate to people? If humans are as sensitive as the guinea pig, the acute lethal dose by ingestion for a 70 kg human would be 42 µg (70 kg × 0.6 µg/kg). If the sensitivity is like the majority of other animals, the acute lethal dose would be between 7,000 and 14,000 µg (7 to 14 mg). Where does the human fit on the scale of sensitivity to TCDD? The tragedy of Seveso has helped provide us with some clues.

On July 10, 1976, a sudden and massive exposure of the entire town of Seveso, Italy, and all of its occupants, both human and animal, to TCDD occurred as a result of an accident in a plant that manufactured trichlorophenol, a chemical used in the production of the antiseptic hexachlorophene. This accident released reaction materials that contained 2 to 3 kg of TCDD (some estimates run as high as 6 to 7 kg). The contamination covered an area of about 700 acres, with the heaviest concentration occurring in a 250-acre area. The unsuspecting population of tens of thousands in the surrounding community knew only that an explosion had occurred at the plant.

Within six days, 11 children were hospitalized with severe skin eruptions eventually diagnosed as chloracne, a clinical symptom of exposure to TCDD. At about the same time, small animals—rabbits, chickens, wild birds— began dying. Then more children developed chloracne and larger animals, such as dogs, sheep, cows, and horses, began dying. In all, there were about 200 cases of chloracne, almost exclusively in children, and over 2,000 animal deaths. Evacuation of the population was begun more than two weeks after the explosion, when soil and grass samples showed that very dangerous levels of TCDD were present in soils closest to the explosion. Some samples were found to contain as much as 5,000 micrograms per square meter.

The number of people involved in the zones around the plant that were judged to have high, moderate, and minimal exposures were approximately 750, 4,700, and 32,000, respectively. All of the women in the high- and moderate-exposure zones who were in their first three months of pregnancy were offered therapeutic abortions because it was known that TCDD is a potent teratogen in animals. Of the approximately 150 women who were in their first trimester at the time of the accident, about 30 accepted the offer despite religious proscriptions against abortion. Among the 120 women who did not choose abortion, about 20 spontaneous abortions occurred, a number not greater than normal.

When news of the accident became public, toxicologists around the world waited with apprehension for the birth of these babies in early 1977. Fortunately, of the 100 or so women who came to term, only 2 bore children with anomalies, well within the normally expected number. One child had an intestinal obstruction and the other a genital malformation, both corrected by surgery.

There is no information on the reasons for the spontaneous abortions; the numbers were overall within normal limits, but 1 or 2 percent higher in the high-exposure zone, a difference that was not statistically significant. Examination of the abortuses from the women choosing therapeutic abortion and of some that spontaneously aborted, although difficult to assess because of the trauma inflicted by the procedure, showed only one case of possible abnormality. That was a case suggestive of Down syndrome, a defect that occurs at the time of conception and thus could not be attributed to the accident.

The people of Seveso were removed from their homes for anywhere from a few months to several years while the government and the industry responsible for the accident tried to decide how to decontaminate their living areas. Homes in the high-contamination area remained contaminated for many years. TCDD penetrated the soil to a depth of almost a foot in the high-exposure zone. Chemical analyses of the soil of Seveso yielded values as high as 20,000 mg TCDD/acre. The lowest level of TCDD found in the 250 acres that were most highly contaminated was about 400 mg/acre. In addition, milk samples taken from cows in the high-contamination zone between three and seven weeks after the accident contained from 1 to 7 ppb TCDD.

The people who moved back to Seveso continue to be exposed to traces of TCDD. Since 1976, and for the rest of their lives, they play their involuntary roles as human guinea pigs. Their vital statistics, their chromosomes, their blood chemistry, their immunologic status, their cancer rates, and their general medical profiles have been and will continue to be of great interest to scientists worldwide. The British medical journal *Lancet* reported in the fall of 1982 that the town and surrounding area had not suffered dioxin poisoning, except for the chloracne cases in children. As late as 2009, there have been no observed chronic adverse effects from the tragedy that befell Seveso.

The final page will not be written until the people of Seveso and their children have lived out their lives. It is an important finding, however, that extremely high exposures to TCDD for a period of more than two weeks' duration to thousands of people of all ages and conditions, particularly pregnant women in the critical first trimester, had essentially no immediate adverse effect. No definite adverse chronic effects have been observed as of the latest report in 2001. The Seveso episode poses a very important question: Why did a chemical that is so extremely toxic to all animals, including the monkey, not cause more immediate damage to a human population? The lack of human toxicity indicates that TCDD may be another chemical for which the monkey is not an appropriate model for humans.

The experience in Seveso tells us that the human being appears to be less susceptible to the adverse effects of TCDD than laboratory and domestic animals. It does not tell us that we should be any less cautious in our regulation of the compound—it is extremely toxic to many other species for which we must also be concerned, and the effects of the compound may be so subtle as to not be distinguished from background in humans.

In 2004, Viktor Yushchenko, a Ukrainian politician, was apparently poisoned with or deliberately took dioxin and had blood levels about 50,000 times higher than normal. He suffered from acute pancreatitis and severe chloracne and still bore the scars from the acne in 2009; he appeared, however, to have had no other adverse events from this massive exposure.

PCBs and Dibenzofurans: Yusho Disease

Polychlorinated biphenyls (PCBs) are chlorinated aromatic hydrocarbon chemicals that have been in commercial use for over 50 years. Because of their physical and chemical properties, PCBs have found wide application as dielectric fluids in capacitors and transformers, heat transfer fluids, plasticizers, and general industrial fluids for engines and pumps. The PCB structure is based on biphenyl, a 12-carbon compound composed of two 6-member rings attached to each other by a carbon-to-carbon bond (see Figure 9-2). PCBs can accept from 1 to 10 chlorine atoms in their structures, 1 chlorine for each of the 10 carbons that are available to interact. There are 209 possible combinations of biphenyl with chlorine. Commercial PCBs are not manufactured as pure compounds but rather as mixtures of chlorinated biphenyls. Their properties are dependent on the fraction of chlorine contained in the formulation, typically between 20 and 70 percent. Some of them are resistant to degradation and persist in the environment for long periods of time.

A wealth of data in the scientific literature demonstrates the toxic effects of PCBs. Acutely, PCBs are classified as having low toxicity by all three routes

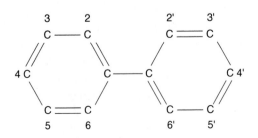

FIGURE 9-2   Biphenyl structure.

of exposure. No toxicity studies have been performed on an individual puri-
fied PCB, but studies of various percentages of chlorine indicate that their
toxicity increases with increasing chlorination. Based on chronic (repeat
dose) animal studies, PCBs also appear to be relatively safe. Evidence in the
scientific literature shows that PCBs, like the majority of the chlorinated
hydrocarbon compounds, produce liver tumors and occasionally liver
cancers in rodent species. The significance of these findings for humans is
unclear because the carcinogenic effect seems to be confined to rodents.
However, using the weight of evidence method for establishing carcinogenicity, PCBs are considered as possible human carcinogens.

There is evidence that at least part of the chronic toxicity of PCBs is due
to impurities that occur in trace amounts in PCB formulations. These impurities, known as dibenzofurans (DFs), are relatives of the DDs discussed earlier.
DFs are also composed of two 6-member carbon rings. However, unlike DDs,
DF rings are joined together by one oxygen bridge and one carbon-to-carbon
bond rather than two oxygen bridges. DFs contain 12 carbons and 1 oxygen
(see Figure 9-1). Like DDs, DFs can accept up to 8 chlorine atoms, but they
can form many more combinations with chlorine than dioxin because of the
differences in the way their 6-member rings are connected. Chlorinated DFs
have not been the subject of as many studies as the chloro DDs. The data
that are available indicate that chloro DFs are similar to chloro DDs in their
mechanism of toxic action but are from 10 to 1,000 times less toxic. For
example, the guinea pig appears to be the animal most sensitive to TCDF
(the chloro DF analogous to TCDD), as it is to TCDD, with an LD_{50} of 5
to $10 \mu g/kg$ whereas other rodents have LD_{50} values greater than $6,000 \mu g/$
kg. Although chloro DFs are less toxic than chloro DDs, they are among the
most highly toxic of synthetic compounds.

A number of unfortunate accidents have involved PCB contamination
of foods. These episodes in which humans have served as involuntary guinea
pigs have provided toxicologists with the clinical symptomatology of subacute and chronic poisoning by PCBs in humans. One of the most extensively
investigated and documented episodes of mass intoxication occurred in
Japan in 1968. Several hundred people in one area of the country became ill
with an unknown disease whose symptoms included acnelike eruptions;
swelling of the upper eyelids; hyperpigmentation of the mucous membranes,
skin, and nails; and discharge from the eyes. The symptoms led to a tentative
diagnosis of chloracne; eventually it was determined that the etiologic agent
was one particular brand of rice oil contaminated with PCBs. The disease
became known as Yusho, which means "rice oil disease."

During the investigation into the cause of the illness, many more hundreds of people became ill with Yusho. The highest concentration of PCBs

found in the rice oil was about 3,000 ppm and the chloro DF content of the PCB contaminant in the rice oil was calculated to be about 0.5 percent, equivalent to 5,000 ppm. Thus, the rice oil contained chloro DF at about 15 ppm. This level of contamination is much higher than that normally found in commercial PCBs, which are usually less than 2 ppm, with the TCDF component less than 0.05 ppm.

In the Yusho episode, the maximum quantity of contaminated oil consumed per person over a three-month period was four to five liters. An investigation of the relationship between Yusho and the ingestion of PCBs led to the estimate that total doses of PCBs greater than 0.5 g over a several-month period produce moderate to severe symptoms of Yusho disease. The maximum total dose of PCBs was calculated to be between 3 and 4 g.

More than 50 percent of the pregnant women who became ill with Yusho gave birth to infants with symptoms of the disease, indicating that PCBs (or the chloro DF contaminant) crossed the placental barrier. However, the possibility that some cases of infant illness were not the result of a direct toxic action on the fetus but rather of nonspecific illness secondary to poisoning in the mother cannot be ruled out. Yusho caused no deaths during the disease outbreak, except possibly for a few cases of stillbirths. No deaths have been attributed to the episode in the years since its occurrence. The people involved in the Yusho episode, like the people of Seveso, are the subjects of medical surveillance for the rest of their lives. However, unlike the Seveso incident, the Yusho episode apparently has caused some cases of chronic illness such as changes in the menstrual cycle and poor vision.

The experience with Yusho cannot be translated directly to occupational exposures because the former occurred with ingestion of PCBs whereas the exposure of working people is usually by inhalation or skin contact. The occurrence of chloracne in occupations using PCBs led to the establishment in 1942—about 13 years after the commercial introduction of the chemical— of a recommended maximum allowable concentration (MAC) for PCB in the workplace of 1 mg per cubic meter of air. About 15 years later the recommended MAC for more highly chlorinated PCBs (54 percent chlorine) was lowered to $0.5 \, mg/m^3$. The $1 \, mg/m^3$ MAC was retained for the less-chlorinated (42 percent) formulations.

Numerous epidemiologic surveys of people exposed for many years to PCBs in occupational settings have revealed no adverse effects attributable to PCB exposure, other than occasional cases of chloracne and possibly diminished liver function. In 1977, the National Institute for Occupational Safety and Health (NIOSH) recommended that the permissible exposure level for PCBs in the workplace be reduced to one thousandth of the current standard, from $1\,mg/m^3$ to $1\,\mu g/m^3$, based on an occasional finding of slightly elevated levels of certain liver enzymes in the blood of a few people exposed to PCBs and on data on liver tumors produced by PCBs in rodents.

In 1979, the Environmental Protection Agency (EPA) prohibited the manufacture of PCBs in the United States and provided for severe restrictions and an orderly phase-out of its uses over a five-year period. This ruling was deemed necessary because of suspected carcinogenicity and persistence in the environment. The full text of this important ruling appeared in the May 31, 1979 issue of the *Federal Register* (44: 31, 514–51, 568). The cost-benefit ratio of the PCB ruling, an important aspect of such regulation usually ignored by the public, was made by the American Council on Science and Health in an article, "PCBs: Is the cure worth the cost?" (May 1986). Despite the ruling by EPA, PCBs will be with us for many years to come.

Many decades of PCB use in transformers, capacitors, industrial machines and pumps, and so forth, has produced a tremendous number of machines and pieces of equipment containing PCBs in use, and that situation will continue for a long time. Eventually they will be replaced by equipment that does not contain PCBs; meanwhile, the old equipment is subject to accident. Because PCBs have a very long half-life in the environment, they have become distributed in quantities of parts per billion or trillion throughout the world.

PCB accidents certainly should not be treated lightly, and the people who clean up such spills should follow work practices that minimize their exposure. But public panic is unwarranted. People who have been accidentally exposed to oils containing PCBs can remove the oils from themselves and their clothing with a good soapy washing. Furniture, automobiles, and other items also can be washed to remove visible signs of contamination. After exposure, individuals should not permit the contamination episode to become a source of worry and the fear that they will develop cancer in 10 or 20 years.

There is no evidence that PCBs contaminated with trace quantities of DFs cause cancer in casually exposed humans. The best of this evidence is from epidemiologic studies of people whose jobs involved daily contact with PCBs. The absence of increased cancer rates in these groups makes it very unlikely that infrequent accidental exposures carry any cancer risk for the general public. It should be noted that the United States has had many cases of PCBs released into the environment, including up to 1.3 million pounds into the Hudson River over a 30-year period. The contaminated river section, which was declared a Superfund site, was still undergoing cleanup operations more than 30 years later, and fishing remained only catch-and-release.

Bophal, India

In 1984, a Union Carbide pesticide plant in Bophal, India, released methyl isocyanate gas into the air. Many people believed that the subsequent deaths (some estimates range to over 25,000 in the following years and up to 3,000 to 4,000 within a short period of time) were due to pesticide exposure. This is not demonstrably false. Methyl isocyanate is a highly toxic compound employed in the process of manufacturing Sevin (carbaryl), a broad-spectrum pesticide used to control beetles, weevils, and other pests without harming vegetation. Sevin compound has been rigorously tested and is a common ingredient used in commercial agriculture and by the gardener.

Assessing the symptoms associated with the release of the methyl isocyanate gas was complicated by the fact that the gas cloud also may have contained any or all of these chemicals:

- Phosgene, a dangerous nerve gas
- Hydrogen cyanide, a lethally toxic gas
- Various chlorine gases
- Carbon monoxide, lethal in high concentrations

Acute symptoms were shortness of breath, burning eyes and throat, and stomach pains with vomiting. The causes of death were listed as pulmonary edema, heart failure, and choking accompanied by necrosis of the kidneys, liver toxicity, and edema of the brain. Additionally, many people were trampled to death as they tried to escape the gas cloud. Stillbirths increased 300 percent, and deaths of newborns doubled. Several hundred thousand people suffered chronic effects from exposure that included:

- Defects on the female reproductive system and an increase in birth defects in children born to exposed females
- Chronic respiratory problems
- Ocular toxicity
- Damage to the nervous system

These toxicities would be predicted from our knowledge of exposure to the chemicals possibly contained in the gas cloud.

Carbaryl can be manufactured by alternate synthetic pathways that use less dangerous materials, but in order to keep production costs down, this plant used the cheaper method employing methyl isocyanate and did not train its workers as to all the dangers surrounding its use and handling. Well into the twenty-first century toxic chemicals remain around the plant in soil and in the well water, as they were never cleaned up; the effects of these chemicals on the population continue. The Bophal accident was a tragedy and is listed as perhaps the worst industrial accident ever to have occurred, but it is unrelated to the toxicity of the pesticide that was being manufactured.

Minamata Disease

First described in 1956 and named after Minamata City, Japan, this form of severe mercury poisoning was caused by the release of methyl mercury from a chemical factory owned by Chisso Corporation. The disease is sometimes referred to as Chisso-Minamata disease. The toxicant was discharged into the water, where it was ingested by fish, which accumulated the mercury. The local population then ate the contaminated fish and suffered the symptoms of severe mercury poisoning, such as vision damage, numbness in the extremities and problems with walking as well as fetal abnormalities, including a high incidence of cerebral palsy. The discharge of methyl mercury waste from this factory continued until 1968. In 1965, there was a second cluster of methyl mercury poisonings due to a release from a factory in Niigata Prefecture, Japan, which was also located along the banks of a river. In 1971, Japan formed an Environmental Protection Agency because of these two environmental disasters along with other problems in Japanese manufacturing safety occurring in the 1960s. The establishment of the Japanese EPA has resulted in the cleaner operation of factories in Japan just as the establishment of the U.S. EPA in 1970 has done in America.

DDT

Dichloro diphenyl trichloroethane (DDT) was one of the first of a host of synthetic organic chemicals developed to fight insects. DDT was originally synthesized in 1874, but it was not until 1939 that its insecticidal properties were discovered in the Swiss laboratories of J. R. Geigy by Paul Müller. The first large-scale use of DDT occurred in January 1944, after several years of worldwide investigation of its effectiveness against insects and its toxicity to humans and animals.

In December 1943, a typhus outbreak among war refugees in Naples threatened to the entire population of that Italian city. Within a month after the outbreak, Allied forces began a DDT dusting program to combat the body louse, the vector of typhus. During January 1944, 1.3 million people were powdered with DDT at the rate of 10 pounds per 150 persons. By the time the dusting program was completed several months later, the typhus outbreak was well under control, the first time in history that this disease had been controlled during winter months.

The great effectiveness of DDT in controlling the insects that transmitted disease led to a decision by the Allied forces that was announced by the British prime minister, Winston Churchill, in a radio broadcast in September 1944: "The excellent DDT powder, which has been fully experimented with and found to yield astonishing results, will henceforth be used on great scale by the British forces in Burma and by the American and Australian forces in the Pacific and India in all theatres."

Thus began the worldwide distribution of DDT. Soldiers and civilians used it as a body and clothing powder to kill lice and ticks. Handfuls of DDT were tossed into rain barrels, cisterns, ponds, and lakes—all containing water that was used for drinking and bathing—to kill malaria mosquito larvae. DDT saved millions of lives from insectborne diseases. It continues to do so in third-world countries, where it is estimated that 4,000 metric tons are produced every year.

When the production of DDT finally exceeded the wartime demands, its use as a weapon against food crop pests began. The American Association of Economic Entomology heralded the extensive agricultural use of DDT that occurred in the postwar years, with a statement in September 1944 saying: "[N]ever before in the history of entomology has a chemical been discovered that offers such promise to mankind for relief from his insect problem as DDT." In 1948, Dr. Müller received the Nobel Prize in Medicine or Physiology for his discovery of DDT. In 1962, DDT was brought to public attention in Rachel Carson's best-selling book, *Silent Spring*. From then until DDT was finally banned for most uses in the United States by EPA in 1972, it was decried as the scourge of our planet. What transformed DDT from Winston Churchill's "excellent powder" into Rachel Carson's "elixir of death"?

The answer lies in the age-old adage that hindsight is better than foresight. It was known that DDT was highly resistant to degradation and that it was stored in the body fat of organisms exposed to it, but the full significance of these properties for its spread throughout the world was beyond the scientific sophistication of the 1940s. In addition, the stimulus that DDT's success gave to the field of synthetic organic chemistry spurred significant improvements in analytical chemical methods. These refinements enabled chemists to detect quantities of DDT (or PCBs, as discussed) in quantities of parts per million, thereby revealing to Ms. Carson and other scientists the extent of DDT's spread in the environment.

The environmental movement received tremendous impetus, if not actual birth, from Ms. Carson's book. Laboratory analyses demonstrated that DDT was virtually omnipresent throughout the world. The decline in populations of wildlife species, such as the brown pelican and peregrine falcon, were attributed to environmental contamination by DDT. A campaign to ban DDT was initiated, and the accompanying publicity gave rise to widespread fears that people, too, were harmed by DDT. The campaign was successful. The use of DDT was banned in the United States within 10 years because the chemical was responsible for wildlife declines.

By 1977, brown pelican and peregrine falcon populations began to increase. In 1989, the Sierra Club announced that the brown pelican, large colonies of which had returned to the California coast, was no longer on the endangered species list. In 2008, the brown pelican was removed from the U.S. endangered species list.

Today, residues of DDT are still found in air, fresh and sea water, fish and other marine organisms, and birds taken from all around the world. The continued presence of DDT in the environment is attributed primarily to its continued use by some third-world and Eastern-bloc countries. In 2008 it was reported that birds nesting in marshes near Chicago contain DDE, one of the metabolites of DDT, at a concentration of 1 ppm—not enough to damage the birds, but certainly indicative of the persistence of DDT and its metabolites in the environment (James Janega, "Bird haven hides traces of poison," *The Chicago Tribune*, May 25, 2008).

Predictions that environmental contamination by DDT would cause an increase in cancer incidence have not been fulfilled. In a 1972 paper, researchers examined the practical application of one risk-estimate formula to DDT exposure:

If one … examines the exposures to DDT which would be "unsafe," accepting the 10^{-6} safety factor for man suggested by Schneiderman (1970) and applying his preferred curve, a serious problem emerges. According to his method, the exposure which should cause tumours in 1 in 100 individuals (TD_1) is a total dose of 30 ng/man/day. In view of the fact that the whole human population of the earth has been ingesting at least 1000 times this quantity of pesticide daily for the past 25 years, one would expect that were DDT indeed carcinogenic for man, and were these safety assumptions based in reality, there ought to have been a massive increase in primary liver cancer among humans in the age group at highest risk, that is, in those who are now between 50 and 70 years old. In effect, the average daily intake of DDT by man over these 25 years would correspond approximately to Schneiderman's TD_{50} and thus, in the population at risk, exposed for at least a quarter of their lives, every eighth person should have developed hepatocarcinoma.[9]

—G. Claus, "Chemical carcinogens in the environment and in the human diet," *Food and Cosmetics Toxicology* 12 (1974): 737.

This example is even more striking when one considers that liver cancer is quite uncommon and that it is often grouped in the catchall category of "All Other Cancers" in compilations of cancer morbidity and mortality statistics. It is estimated that liver cancer represents about 1 percent of all cancers in the United States and that many of these (>30%) are associated with cirrhosis due to alcohol consumption. There is no evidence that DDT is a human carcinogen, although it is currently listed as a probable carcinogen based on animal studies. The controversy over its human carcinogenic potential continues.

Lest one think that DDT is the only pesticide found as a residue in the environment, *The Chicago Tribune* in an article entitled "Pesticides in your peaches" (August 12, 2009), reported that more than 50 pesticides can be found on produce, including peaches, apples, carrots, lettuce, and celery. This kind of news is certain to frighten some people who will equate "pesticide" with "poison." By now readers know that they should do more research and try to determine the amount of pesticide present, which ones, and the actual danger that the residues impart. The article then revealed that peaches from sources such as those certified organic and those from a farmer's market contained very low levels of various approved pesticides—many times less than what EPA permits. One pesticide was found at a concentration of 157 ppb, which is 20-fold less than the EPA allowance for organic produce. Because we have refined our testing methods considerably, it appears that we will be able to detect smaller and smaller amounts of all kinds of chemicals, and the public either will be scared to death or will learn how to assess the risk associated with the data. We address this topic in Chapter 11.

CONSUMER PRODUCTS

Metals

We have discussed metals in several places throughout the book, but there are more issues surrounding metals than just those seen with lead and mercury. Metals such as sodium, potassium, copper, and zinc are essential for life. However, like all other elements and compounds, in higher amounts they can cause toxicity. Metals such as silver and gold have medicinal use. Metals such as lead and cadmium are toxic to humans in almost every amount.

One of the problems with metals is that they are ubiquitous, found almost everywhere in nature as well in a multitude of manufactured products. Therefore, we are exposed to metals on a daily basis, either knowingly, as in when we take our daily vitamin pill or eat processed foods, or unknowingly, when we use products containing metals that can leach out. A few examples of inadvertent exposure to metals are discussed next.

Lead and Cadmium

In 2009, there were multiple reports in the media concerning levels of lead in toys and children's jewelry imported from China. Items were then tested for lead content and either were banned from import or recalled from the market if already in the United States because lead content is regulated by law. In order to remove the lead from these toys, some manufacturers substituted cadmium for the lead. Although lead is a serious toxicant, leading to impaired function when children are exposed to it during neuronal development, cadmium can be an acute toxicant, leading to liver and kidney damage as well as neuronal damage similar to that seen with exposure to lead. Cadmium is also a carcinogen. According to a CBS news story (January 11, 2010), concentrations of cadmium in some toys tested were greater than 80 percent by weight. At the time these items were imported, there were no legal restrictions on cadmium content in jewelry, although there are restrictions on the amount of cadmium permissible in painted toys. It appears that cadmium was used in place of lead because the price for cadmium was down and lead use had been severely restricted. This decision may have put children into a greater danger than the original composition of the items.

Zinc

Zinc is an essential metal that is involved in many enzymatic reactions in the body, including protein synthesis, functioning of the immune system, and normal cell division. Zinc is found in many foods; a healthy diet should provide the suggested daily amount of zinc of approximately 10 mg/day. A multivitamin supplement usually provides 100 percent of the recommended daily intake. Some older adults can suffer the effects of zinc deficiency—hair loss, diarrhea, weight loss, impotence, and taste abnormalities are the most common—if their diets are low in meats and if no dietary supplements are taken. However, this deficiency is relatively rare in the United States among people who eat meat; it is somewhat more common among vegetarians and those not able to efficiently absorb zinc from the gastrointestinal tract (National Institutes of Health, Office of Dietary Supplements, Zinc Fact

Sheet). The upper limit for adult intake of zinc is about 40 mg/day with side effects of overdose, including nausea and vomiting, after acute exposure and reduced immune function with chronic exposure. Additionally, too much zinc can interfere with copper and iron in the body, both of which are also essential for certain enzyme systems to function properly.

Zinc has been suggested to be useful in shortening the signs, symptoms, and duration of the common cold and is sold over the counter in lozenges, gels, and nasal spray formulations, although there is some controversy in the medical community as to whether this effect is real. Several articles reviewing all the studies have come to opposite conclusions.

Unlike metals such as lead and cadmium, it would seem that zinc in various products should be of no concern to the public. However, in 2010, GlaxoSmithKline, the maker of Poligrip® denture cream, stated that it would not include zinc in any of its formulations. This action was a direct result of several lawsuits which claimed that excessive use of these products led to neurological damage and various blood problems, symptoms that are claimed to be due to chronic zinc exposure. (M. Perrone and C. Anderson, "Glaxo to remove zinc from denture cream," *AT&T News*, February 18, 2010). Poligrip contained 38 mg zinc/oz, and some people were using a tube every week, which led to about a 45-fold increase over the recommended daily amount.

Plastics

In our industrial society, plastics are almost as ubiquitous as metals. It is difficult to shop in a store in America without coming home with plastic, usually as a food wrap, a carrying bag, or a container. Although plastics are a great convenience and have led to a reduction in food contamination, there are also problems with plastics which come into contact with our food, drink, and medical supplies. If the precautionary principle were in effect in the United States, the chemical classes presented next would be restricted. Since they are not, it behooves readers to make their own decisions about exposures to these chemicals.

Phthalates. Plastics tend to be stiff materials. In order to make them flexible, phthalates, the esters of phthalic acid, are added. Before phthalates came into use, compounds such as castor oil and camphor were used. Phthalates were much more consumer-friendly since they did not leave an oily taste or have a bad smell. However, they easily leach out of the plastic material and can be ingested. Additionally, certain capsules for medicines contain phthalates, which are then absorbed into the bloodstream. There is a whole family of phthalates, each one of which has slightly different chemical and physical properties.

According to the Centers for Disease Control and Protection, most Americans have phthalate metabolites in their urine. Phthalates belong to a group of chemicals that are known as *endocrine disruptors*, that is, compounds that can upset the normal function of the endocrine system by acting like the body's own hormones. Many studies of phthalates and other endocrine disruptors such as polychlorinated biphenyls, Bisphenyl A (to be discussed below), and some pesticides such as atrazine and DDT have shown effects on animal reproduction in toxicology studies. Because many of these compounds are in such widespread use, their direct effects on human reproduction are hard to determine. In one study by Sathyanarayana et al., 163 infants were measured for urinary phthalate metabolites (S. Sathyanarayana et al., "Baby care products: possible sources of infant phthalate exposure," *Pediatrics* 121 (2008): e260–e268). More than 80 percent had phthalate metabolites in their urine, and the concentration was related to the recent use of baby lotions and shampoos. Although there is no doubt that infants, children, and adults are exposed to phthalates repeatedly, there is a question as to the effects of this exposure.

To date only one study in humans seems to show an effect. In that study, S. H. Swan and other investigators measured the distance from the anus to the male genitalia in 134 boys and found an association with phthalate metabolite concentrations in the urine and a shortening of the anogenital distance (S. H. Swan et al., "Decrease in anogenital distance among male infants with prenatal phthalate exposure," *Environmental Health Perspectives* 113, no. 8 (2005): 1056–1061). They hypothesize that this shortening is a marker for disrupted reproductive development. However, other reproductive scientists have taken issue with the conclusion since they feel that the anogenital distance is not a well-understood marker for reproductive development in humans and that the study size is too small (see J. R. Barrett, "Phthalates and baby boys: Potential disruption of human genital development," *Environmental Health Perspectives* 113, no. 8 (2005): A542.35). And so the discussion goes on as to the effects of phthalates in humans. A prudent consumer might wish to reduce exposure to these compounds unless some substantial evidence of their lack of effect is found.

Bisphenyl A

Bisphenyl A (BPA) is a component of many plastic materials, particularly polycarbonate plastics, which are hard materials commonly used in reusable water bottles, equipment for various sports, and medical devices. Several studies have found effects of BPA on animal reproduction as well as an implication that it may be involved in obesity. It is classed as an endocrine disruptor as well as a compound that alters thyroid function. BPA in the water has been shown to affect the development and reproductive ability of aquatic animals. To date, no data show that it is a carcinogen, which may be because no definitive testing has been done. There is, however, evidence that exposure to BPA during gestation leads to an increase in breast cancer in rats (see C. Brisken, "Endocrine disruptors and breast cancer," *CHIMIA International Journal for Chemistry* 62 (2008): 406–409).

A study was published in 2010 in which pregnant rats and their offspring were given doses of BPA up to 410 mg/kg/day (2250 ppm) each day from the first day of pregnancy through 21 days of lactation, the normal nursing period for rats (D. G. Stump et al., "Developmental neurotoxicicity study of dietary Bisphenol A in Sprague-Dawley rats," *Toxicological Sciences* 115 (2010): 167–182). There were no effects on a wide variety of neurological parameters monitored in the study at any dose. There was, however, a dose-related decrease in both maternal and offspring weight at doses higher than 13.1 mg/kg/day (75 ppm). Although this study was sponsored by an industry group— the Polycarbonate/BPA Global Group—it was performed at a reputable contract laboratory under the conditions of Good Laboratory Practices and published in a reputable scientific journal after peer review, indicating that the results are credible.

Some countries have banned the use of BPA while others allow it. The situation in the United States is state and city specific, with some municipalities banning the material while others allowing it. Several manufacturers have stopped making BPA; however, it is still legal in many places. In 2010, EPA put BPA on its list of chemicals of concern, and the National Institutes of Health has $30 million in grant money available to study its health effects. It is hoped that new data will be available in the next several years so that safety standards for BPA can be determined. In the meantime, minimizing exposure to this compound seems wise.

INDOOR AIR POLLUTION

Indoor air pollution has become a common problem for modern society. These adverse health effects are referred to as tight building syndrome and sick building syndrome. Indoor air pollution is included in this section because it represents another complex environmental health problem that affects a large number of people. Beginning in the early 1970s, increasing numbers of complaints and reports of illness from office workers called attention to this new health problem associated with indoor air quality. It differs from typical occupational health problems in that it is caused most often not by one or two airborne contaminants but usually by a large mixture of trace quantities of pollutants, none of which alone appears capable of causing illness.

Illnesses associated with office buildings were recognized many years before indoor air pollution was identified as the cause. The Bateson Building, a State of California office building in Sacramento that opened its doors to its first occupants in 1981, is an early example in which the relationship between indoor air pollution and illness was readily apparent. This new building was the realization of a much-publicized effort to create a state building that would serve as a model for the ultimate in energy conservation. Within a year, the building was the target of a class action lawsuit on behalf of the 1,200 employees. Their suit was based on the claim that working in the building made people ill. A survey by an employee organization showed that 80 percent of the employees on one floor had one or more symptoms that they attributed to the building environment. These symptoms included various ailments of the upper and lower respiratory tracts, nausea, itching or burning eyes, sinus problems, skin irritation, dizziness, and fainting spells. An engineering survey concluded that the ventilation system was providing inadequate fresh air circulation.

The problems found in buildings such as the Bateson Building are caused most often by poor ventilation that leads to an accumulation of a wide range of air pollutants. Most frequently, investigations show that indoor air pollution is composed of airborne gases, vapors, particles, odors, and exhaled air that have been released inside the building by structural materials, clothing, people, office products, office furnishings, and (in the past) smoking. Less frequently noted contaminants are chemicals and solvents from office photocopying equipment, products of combustion from stoves and furnaces, and growths of molds and other microorganisms in wet areas of the building or its ventilation system.

Occasionally polluted air enters the fresh-air intakes of the ventilation system from outside the building, especially from automobile exhaust, cooling tower mists, products of combustion from nearby smokestacks, and air pollutants from nearby industry. The majority of problems, however, are caused by the buildup of airborne contaminants as the result of building designs that restrict the intake of fresh air to conserve energy. In some cases, only parts of the building are poorly ventilated. In other cases, the ventilation system is adequate but poorly maintained or operated.

The problems caused by microbial contamination deserve further mention because of the serious illnesses and deaths that sometimes occur. One example, the outbreak of Legionnaires' disease in a Philadelphia building in 1976, resulted in 182 cases and 29 deaths due to acute respiratory infections. The cause was traced to airborne *Legionnella* bacteria. This is an example of a general class of an often-serious but infrequent indoor air pollution problem: Bacteria, fungi, or other microorganisms contaminate and then multiply in cooling tower water, in stagnant wet areas inside ventilation systems, or in wet carpets or furnishings. Under the right conditions, these organisms will become airborne and be dispersed by the building's ventilation system. The resulting effects depend on the organism involved.

Prevention depends on the elimination of potential sources of microbial contamination of building intake air and control of microbial growth within the building by reducing or eliminating stagnant water accumulation in ventilation systems or on carpets and furnishings. In addition to pathogen-borne diseases, a variety of disorders, such as hypersensitivity pneumonitis, humidifier fever, allergic rhinitis, and conjunctivitis, have been reported, due to indoor air contamination.

In 446 episodes of tight building syndrome investigated by the NIOSH, the primary causes were classified in this way (*Indoor Air Quality*, September 1989):

Inadequate ventilation	52%
Contamination from inside the building	17%
Contamination from outside the building	11%
Microbiological contamination	5%
Contamination from the building fabric	3%
Unknown	12%

More recent investigations have revealed somewhat similar profiles. Anyone who has been affected by indoor air pollution or who wants to learn how to conduct an investigation of indoor air quality should look at the NIOSH information on Indoor Air Quality (www.cdc.gov/NIOSH/), where handy checklists are available for use in surveying a home or office.

What are the pollutants in indoor air? Studies have identified many of the contaminants as chemicals originating from the occupants, furniture, structural materials, cleaning and polishing materials, and office machinery. Many other chemicals in indoor air are not known and may never be identified. The composition of indoor air pollution is immensely variable. The fact that inadequately ventilated rooms made people ill long before the advent of the petrochemical industry suggests that stoves and furnaces, pathogens (viruses, bacteria, molds), natural allergens (pollens, animal dander, etc.), tobacco smoke, and even the products of human metabolism exhaled from the lungs or excreted in the sweat are significant factors. The suggestion has even been made that the buildup of radon, a naturally occurring radioactive gas given off by soils and rocks and by building materials made from them, in unventilated rooms also contributes to illness. Of all of the chemicals from human sources, carbon monoxide has the greatest potential for causing illness or even death, because it has the greatest potential for achieving dangerous air concentrations from inadequate ventilation of combustion sources.

The use of synthetic materials in construction and furniture and the use of organic chemicals in office supplies and equipment have added to the burden of indoor air pollution. Probably the one chemical that has caused people the most difficulty is formaldehyde, a simple organic chemical containing only one carbon atom. It is widely distributed in nature and is present in all of us as a product of the metabolism of one-carbon biochemicals. It is also used extensively in the plastics and polymer industry.

As a gas, formaldehyde is extremely irritating and is intolerable in concentrations greater than a few parts per million in air. Preserving solutions for biologic and pathologic materials are usually 3.7 percent formaldehyde. Solutions of formaldehyde are irritating to skin, eyes, and mucous membranes. Formaldehyde is a strong sensitizer, and the usual symptoms that people develop from exposure to formaldehyde vapors are those of an allergic nature, including skin rashes, headache, puffy eyes, and respiratory problems. It has been estimated that perhaps 10 percent of the population is allergic to formaldehyde. The people at greatest risk of adverse effect from exposure to formaldehyde are those who work with concentrated solutions—chemical workers, undertakers, pathologists, and other scientists who work with preserved biological specimens. Their exposure is many orders of magnitude greater than that which results from indoor air pollution. However, given that carpeting, drapes, and furniture may all contain formaldehyde that will degas, the contribution of formaldehyde to indoor pollution should not be minimized.

The carcinogenicity of formaldehyde has been given a great deal of attention; thus, any mention of the effects of exposure would be incomplete without reference to it:

> Formaldehyde gas is carcinogenic for rats and probably for mice, producing nasal tumors after inhalation. Limited experiments in Syrian hamsters have not demonstrated carcinogenicity. In rats, the carcinogenic response appears nonlinear, being disproportionately higher at higher concentrations.[20]
>
> —"Occupational Exposure to Formaldehyde: Final Rule,"
> *Federal Register* 52, no. 233 (December 4, 1987): 46, 205.

Animal studies give no evidence that formaldehyde causes cancer in any site other than the nasal passages. Some epidemiological studies of occupational groups of humans show no association between exposure to formaldehyde and cancer incidence. Others show a weak association. Therefore, although formaldehyde remains a suspected human carcinogen, it is important for the public to recognize that the concentrations encountered in indoor air pollution are much smaller than the occupational exposures that are of genuine concern. The allergenic properties of formaldehyde are of much greater significance to the general public than its potential carcinogenicity. A review of the extensive data on the health effects of formaldehyde can be found on OSHA's Web site (www.osha.gov/SLTC/formaldehyde/index.html).

Indoor air pollution provides an excellent illustration of two principles that are often exhibited during the implementation of advances in science and technology. The first is that the solution of one problem often creates another, possibly more severe, problem. The second is that, throughout history, the figurative wheel keeps being reinvented. In accordance with the first principle, indoor air pollution is the problem created by the attempt to solve the problem of the high cost and shortage of energy. In order to reduce the expenditure of energy necessary to provide a comfortable indoor air environment, buildings are made more airtight to keep the warmed or cooled indoor air from escaping. Fuel is saved, and people are made ill.

In the second principle, the solution to the problem of how to keep people from becoming ill in modern, fuel-efficient buildings may be symbolized by the proverbial wheel. There is a massive effort under way to solve the problem, when the solution is already at hand—the wheel has already been invented. Many years ago ventilation engineering technology developed all of the formulas used to calculate the proper number of air exchanges required for healthful air quality. Additionally, houseplants can act to remove pollution. The National Aeronautics and Space Administration has a list of plants that have been shown to be helpful in removing formaldehyde and benzene (www.nasa.gov/), including ivy, gerbera daisies, and peace lilies.

More research may be needed to devise methods for conserving energy while providing for adequate air exchange, but no further research is needed to prevent adverse effects from indoor air pollution.

WATER POLLUTION

Water quality affects everyone on Earth every day. Many parts of the world do not have a dependable supply of fresh water of good quality. Fortunately, in the United States, we can drink water out of any tap without fear of contaminants—or can we? Let's discuss how consumers can determine the purity of their own drinking water.

If you get your water from a municipality, you should receive a report at least annually about the water coming out of your tap. Evanston, Illinois, where one of the book's authors lives, puts out a water quality report each year and distributes it to all homes. This report lists the substances that are measured—such as fluoride, sodium, lead, copper, bacteria, radium, and others—in parts per million or billion of each substance. It also lists the levels allowable by law for comparison. A review of this report shows at a glance whether the water supply meets federal standards. In addition, Evanston now tests for pharmaceutical and personal care products in the water since some of these products were found in Lake Michigan water from which the city draws its supply. Testing at the water station showed:

Nicotine	0.007 ppb
Cotinine (a metabolite of nicotine)	0.003 ppb
Gemfibrozil (an anticholesterol drug)	0.0010 ppb

Many other products, such as aspirin, caffeine, and DEET, were tested for but not found. A risk assessment was performed to assess for effects of the gemfibrozil concentration. It was found that someone would need to drink two quarts of water each day for 16,000 years in order to ingest the amount in one recommended daily dose of gemfibrozil. Based on all these findings, it appears that the Evanston water supply is suitable for daily consumption.

We are not all as lucky as the citizens of Evanston. For example, another suburb of Chicago—Crestwood, IL—used a well to supply its residents with drinking water. The Illinois EPA prohibited the use of this well for drinking water after volatile organic compounds, such as vinyl chloride, were found to have contaminated it. The agency required this city to purchase its water from the Lake Michigan distributors. However, for about 10 years, that city continued to distribute well water to its citizens in violation of the order. These people were exposed to potentially toxic water every day during that time. A risk assessment for exposure to the contaminants will be difficult, as sampling the water today may not accurately reflect the level of the chemicals during the exposure period.

PHARMACEUTICALS

It was mentioned in the preface that many people are surprised to find out that their medicines, whether prescribed or purchased over the counter, can (and usually do) have side effects—that is, toxicity. This should not be at all surprising since drugs are chemicals, and, as we have discussed, all chemicals have toxicity at a high enough dose. Of course, with medicines, the separation between the therapeutic effect (good) and the toxic effect (bad) should be sufficient to protect most people from experiencing adverse events (toxicity). But because of genetic variability, the Therapeutic Index— the ratio between the effective dose and the toxic dose—varies from person to person for each of the possible side effects that a drug can produce. Let's look at some examples of prescription drugs that have been recalled during recent years to see how the Therapeutic Index shifted lower as a more diverse population took these products for longer periods of time.

Fen-Phen

This combination of two approved drugs, fenfluramine and phentermine, was never approved by the the Food and Drug Administration (FDA), but under U.S. law, any physician can prescribe approved drugs according to his or her medical expertise. Phentermine was approved in 1959 for the short-term treatment of obesity. Its side effects were considered mild and consisted of dry mouth, insomnia, constipation, and the like. Fenfluramine, which is related to amphetamines, was approved in 1973 as a controlled substance, also for the treatment of obesity. Its side effects included stimulation of the nervous system with dizziness, agitation, constipation, and heart palpitations. Both of these drugs were of limited value in a weight-loss program, and their use was not widespread. However, starting in the 1990s, physicians decided to coadminister them for a greater effect. And it worked! People were able to lose more weight and therefore continued taking the combination. Everyone was happy until a number of these patients began to exhibit heart valve problems. It turns out that heart valves have serotonin receptors (a fact that was not appreciated before this time) and that fenfluramine was metabolized to an active metabolite that stimulated these receptors, leading to valve damage. It has been estimated that approximately 30 percent of the people who took fen-phen had some valvular damage, although some of these cases resolved after discontinuation of the medicine. Other people required valve replacement surgery. The company that marketed fenfluramine ended up paying a large settlement as a result of a class-action lawsuit. Fenfluramine was removed from the U.S. market in 1997, and eventually this drug was banned worldwide. Phentermine is still available.

Why was this side effect not discovered earlier? Clinical trials of fenfluramine involved only a small number of patients over a relatively short time.

As it was not a very helpful weight-loss medication, its use after approval was relatively small. Because the drug had unpleasant side effects and it was not very useful, patients did not take it for very long. If there had been patients with valve issues, there would have only been a few, and their problems would have been hard to separate out from background occurrences. However, in combination with phentermine, the drug became much more effective. People were willing to put up with the side effects because they were losing weight, so they took the drug for longer periods of time. Eventually, the heart valve damage became common enough that a risk assessment was able to pinpoint the drug combination as the causative factor. Additional laboratory research was performed to determine which drug in the combination was to blame and eventually what the mechanism of action was to cause the damage.

Vioxx

A few words about nonsteroidal anti-inflammatory drugs—NSAIDs for short—which include over-the-counter pain remedies such as ibuprofen and naproxen, but not acetaminophen, as well as many prescription products. These compounds work by blocking an enzyme called cyclooxygenase (COX), of which there are COX 1 and COX 2 types. All of these over-the-counter products tend to block primarily COX 1 with COX 2 blocked to varying degrees. The compounds produce side effects such as gastrointestinal (GI) upset, including ulcers, and kidney damage. (Aspirin is usually classified as an NSAID although recent investigations have shown that it has the ability to alter the COX 2 enzyme and, by changing the structure of COX, has the ability to be cardioprotective. Aspirin is routinely included in the pharmaceutical armamentarium for this reason.)

While researching this enzyme system, companies discovered how to make compounds that were more selective for COX2 and thus appeared to be GI sparing, which is of great importance since many older people take these products for relief of the pain of arthritis, and they are more prone to the GI side effects. Protecting against stomach upset and ulcers seemed like a good thing to do, and several companies produced COX2 inhibitors such as the prescription medicine Vioxx (rofecoxib).

Vioxx was approved in the United States in 1999, and many millions of prescriptions were filled. The drug did have a better GI profile than the COX1 inhibitors, although it was not completely free of GI problems, and it was just as effective as ibuprofen and others in ameliorating pain. However, in 2004, Vioxx was pulled from the U.S. market after studies showed that patients taking Vioxx had a fourfold increase in myocardial infarction than patients taking a COX1 inhibitor for one year. Although the increase in incidence was from 0.1 to 0.4 percent, which still seems like a small risk, the drug was removed because heart attack is a very serious side effect, the cardiac problems began as early as two months after starting the drug, another study showed increase in stroke, and other products to treat the pain of arthritis and other conditions are available.

What went wrong here? It turns out that the findings with Vioxx triggered additional studies with other NSAIDs, which indicated that the COX1 inhibitors themselves may also contribute to heart problems. Aspirin may be the exception, but, unfortunately, it has the worst GI profile of any of the NSAIDs. Another COX2 product is on the market in the United States—Celebrex—and it has stayed on the market because its effect on the heart is not as severe as that of Vioxx. It turns out that Celebrex is not as specific for the COX2 enzyme as Vioxx, which may be the reason the heart is somewhat spared. The answer to our question seems to be that if a drug inhibits COX1, there are GI side effects; if it inhibits COX2, there are heart side effects; if it is a mixed inhibitor, which most NSAIDs are to some extent, then the type of problem most prevalent depends on which COX is blocked more.

Why were these findings not seen in the drug trials that were required before approval? Because heart problems were an unexpected finding, investigators apparently were not looking for them very hard. In retrospect, there were signs of heart issues in the clinical trials, but they were dismissed or underreported. During many legal battles to date, evidence has been presented to show that the company, or at least some of its investigators, may have hidden data. The company has paid out billions of dollars in settlements of these lawsuits.

Thalidomide

Thalidomide was approved in 1957 in many countries but not the United States. It was an effective sedative that also had antiemetic properties. Because of its effect on nausea, it was prescribed to pregnant women to prevent morning sickness. By 1961, it was clear that thalidomide was a strong teratogen if administered during the time when the limbs of the fetus were developing and that defects such as extra limbs or malformed limbs were produced. The most common defect was phocomelia, in which the long bone of the legs or arms is not formed and the hand or foot is attached directly to the upper limb or shoulder. It is estimated that about 20,000 birth defects worldwide can be attributed to thalidomide, including over 60 in the United States, where the drug was undergoing clinical trials. Thalidomide was recalled worldwide in 1961.

How could this drug be approved in the first place? In 1957, the requirements for testing for teratogenicity were considerably less stringent than they are today. In addition, at that time, our perception was that the placental barrier prohibited the passage of xenobiotics. We now know that this is not true. A teratology study performed in rats before registration of thalidomide did not show any adverse effects. After the recall, a study done in rabbits and monkeys produced phocomelia. So the wrong species was used for the toxicology study. Could this happen today? It is possible but less likely. The International Committee on Harmonisation guidance for teratology testing requires the use of a rodent, usually the rat, as well as a nonrodent, usually the rabbit. In some cases where the known user of the product will be pregnant women, the monkey is also used as the teratology model.

Thalidomide and its analogs have now found a place in the treatment of leprosy and multiple myeloma. They have side effects that are not pleasant, but because of their effectiveness, and as long as they are not taken by pregnant women, they have become useful for these serious diseases.

CHAPTER 10

EPIDEMIOLOGY

Epidemiology is the study of the occurrence and movement of diseases in groups of people and the investigation of the relationships of these diseases to causative factors. The purpose of epidemiology is to describe the symptoms of a particular disease and to discover the chain of association between the disease and its etiological (causative) agent. Along with logical thinking and systematic documentation, epidemiology makes use of a number of scientific specialties in addition to toxicology, such as clinical medicine, pathology, biochemistry, and bacteriology, to collect and evaluate data on health effects. Biostatistical methods are used to look for and evaluate associations between effects and possible causes.

The science of epidemiology arose from an interest in learning the cause of the epidemics of diseases, such as bubonic plague, typhus, and cholera, that swept through Europe and Asia prior to the end of the nineteenth century, leaving huge death tolls in their wake. These diseases had been present since before recorded history, but until relatively recently, they were considered to be acts of God and random in their outbreaks, with no apparent physical cause. Early epidemiologists systematically observed and recorded exactly who contracted a particular disease and who did not, where, when, and under what conditions, and thus were able to obtain sufficient information about cause and effect to prevent some epidemics despite the fact that the specific etiologic agent was not even suspected, much less identified.

ORIGINS OF MODERN EPIDEMIOLOGY

The English physician John Snow is one of the pioneers of modern epidemiology. Snow studied outbreaks of cholera in London during the middle of the nineteenth century. At that time, cholera was a common disease that regularly devastated large populations throughout Europe and Asia. The disease seemed to spread from person to person, but how or why was a

The Dose Makes the Poison: A Plain Language Guide to Toxicology, Third Edition.
By Patricia Frank and M. Alice Ottoboni
© 2011 Patricia Frank and M. Alice Ottoboni. Published 2011 by John Wiley & Sons, Inc.

mystery. Snow recognized that cholera was associated with poverty, over-crowding, refuse, and filth. He also suspected that the distribution of cholera cases in London was somehow associated with the source of drinking water. The frequency of cases seemed to increase with increasing sewage contamination of the water supply.

When an outbreak of cholera occurred in 1848 in a small section of London known as the Golden Square, Snow systematically identified and then plotted each case of cholera on a spot map of the area. This map showed that most of the cholera cases were centered around one water pump known as the Broad Street pump. Further investigation of cholera cases among the population living in Golden Square at the time also showed that two groups had a very low incidence of the disease. One group was a workhouse population of over 500 inhabitants that used water from the workhouse well rather than the Broad Street pump; the other was a population of 70 brewery workers. In checking their drinking water source, Dr. Snow learned that they most likely drank no water at all from the surrounding community because they were provided with free ale! In addition, the brewery had a deep well of its own. Dr. Snow went on to study the number of cases of cholera among the customers of two competing water companies that drew water from different areas of the Thames River. Again his results showed that water polluted with sewage was associated with cholera.

Despite these careful studies, Dr. Snow did not conclude that cholera was transmitted by a pathogen in the water since the concept that microorganisms could transmit disease had not yet been formulated. However, he did show a strong association between water contaminated with sewage and the incidence of cholera. This information alone was extremely important because it gave direction to practical preventive programs and further studies aimed at finding the causative agent. In fact, based on Dr. Snow's investigations, the development of sanitary procedures to prevent contamination of drinking water by sewage virtually eliminated cholera in London. However, the final proof of the cause-effect relationship did not come until years later, when the infectious organism that caused the disease was isolated and identified.

Epidemiology is a rigorous discipline that relies heavily on biostatistical methodology. A competent epidemiologist must not only be well grounded in medical, biological, and statistical sciences but must also be an imaginative and curious detective. The latter qualities are essential to uncover and properly weigh all of the many factors that might be associated with the problem under investigation. A great book that gives many examples of epidemiological studies and has some very nice graphics is *Investigating Disease Patterns: The Science of Epidemiology* (1995) by Paul D. Stolley and Tamar Lasky.

EPIDEMIOLOGY OF NONINFECTIOUS DISEASES

Before proceeding, it is important to dispel the common impression that the biostatistical component of an epidemiologic study can reveal the cause of a disease. This misunderstanding has led to a great deal of public confusion and dissatisfaction with epidemiologic investigations of cancer and birth defect clusters. For example, in a community where a cluster exists, community fatality and/or illness records are sifted for possible associations with environmental factors, such as water source, pesticide use, or socioeconomic status. These investigations may or may not indicate an association. If there is no statistical association, there is almost surely no cause-effect relationship. A weak association may be suggestive but cannot be used immediately for preventive purposes. An association is most useful when it is very strong—that is, when a disease has almost no occurrence in a non-exposed group and a high occurrence in an exposed group. However, association does not necessarily imply a cause-effect relationship; such a relationship can be proven only when all of the components that make up a valid and complete epidemiologic study support it. These components include accurate diagnosis, biostatistically significant differences in comparable population groups, identification of an etiological agent, and evidence of exposure. The key point is that while biostatistical studies may contribute valuable information, they are only one step in the epidemiologic process. People must recognize that the biostatistical component of a study can demonstrate *associations* but not *causation*.

Epidemiology originated a century ago in the study of infectious diseases when death by infection was commonplace. Today, noninfectious diseases, such as lung cancer and heart disease, are important causes of disability and death. These diseases are associated with lifestyle patterns, nutritional factors, age, and chemicals. Experience has shown that epidemiologic investigations of such diseases are much more difficult than investigations of infectious diseases for many reasons. Early symptoms are often silent or vague, latency periods usually are measured in years rather than days, and the diseases in question tend to have a normal rate of occurrence throughout the general population. As a result, associations are not readily apparent.

Extensive use of biostatistics often is required to learn whether the rate of disease in the group in question is higher than the normal background level and whether the disease is associated with any common factors within the group. Unless biostatistical studies yield a significant association, further steps in the epidemiologic investigative process usually are not taken. Unfortunately, the popular press often cites weak or merely suggestive associations as proof of a cause-effect relationship.Epidemiology neverthe-less provides a valuable and often essential adjunct to investigations of the beneficial or adverse effects of exposure to chemicals in human populations. These may be groups of people who work at the same occupation and thus have similar exposure to an industrial chemical, or groups who live in the same community and thus share similar exposure to some environmental contaminant in their air or water supply.

One approach is to compare two groups of people that are as alike as possible except for their exposure to a specific chemical or their incidence of a specific disease. This method uses what is known as matched populations, where either differences in exposure are compared to look for differences in the incidence of disease or differences in incidence of the disease are com-pared to look for differences in exposure.

Epidemiologic studies may be retrospective or prospective. Retrospective studies look back on populations exposed in the past; prospective studies look forward to populations that will be exposed in the future. Retrospective studies suffer from the fact that critical data (what dose of chemical did each person receive and for how long a period of time) often are not available or cannot be retrieved. Such data must be estimated, a procedure fraught with uncertainty and error.

In prospective studies, plans are made in advance to record levels and times of exposure, health status prior to exposure, and so forth. Thus, prop-erly designed prospective studies usually yield valuable information on chronic effects of chemicals. A number of major companies in the chemical industry have instituted programs of prospective epidemiology among their employees for all new chemicals they manufacture and for some of the older ones that have come under suspicion as a cause of health problems. The major deficiency of prospective investigations is that answers are not imme-diately forthcoming. In some cases, it may take many years to evaluate the significance of chronic exposure to a chemical on health. As mentioned previ-ously, it is important to recognize that association is the key word. Statistical analyses can reveal whether two factors are related and, if so, how strongly, but no matter how strong the association, the relationship is not necessarily a causal one.

For example, several decades ago, before it became fashionable for women to wear pants or slacks, an almost perfect association could be demonstrated between the wearing of skirts and the incidence of breast cancer. Another example of a strong correlation between unrelated occur-rences is the decline in the stork population in Europe accompanied by a decline in the human birth rate. Despite the very strong associations, no one would claim that wearing skirts causes breast cancer or that storks bring babies.

These amusing examples do not in any way lessen the great value of statistical analyses in the study of the relationship between disease and exposure to chemicals. However, they do serve as a warning that the public should be very wary of any epidemiological study claiming to prove a cause-effect relationship based on statistics alone.

Koch's Postulates

If a study does show an association between an environmental contaminant and a disease cluster, further work is required to demonstrate a cause-effect connection. The model for determining cause-effect relationship for infectious diseases was first applied in 1876 by the German physician Robert Koch. These criteria, known as Koch's postulates, may be summarized in this way:

1. The organism responsible for the disease must be present in every case of the disease.
2. The organism must be able to be isolated from the patients and grown in pure culture.
3. The organism grown in pure culture must be able to cause the disease in a healthy host.
4. The organism must be recovered again from the experimentally infected host.

Koch's postulates cannot be applied directly to environmental chemicals because of the differences between microorganisms and chemicals in how they cause diseases. However, the postulates are very useful in helping to illustrate the new and difficult problems faced by investigators who study the causes of noninfectious diseases.

Postulate 1: The organism responsible for the disease must be present in every case of the disease. With environmental chemicals, all people with the disease in question may not have been exposed to the suspected chemical. Many of the diseases caused by chemicals have a normal background incidence in the general population. A good example is leukemia, known to be caused by ionizing radiation and by benzene but which also has a measurable background incidence in the general population. Thus, the investigator searching for causes of leukemia in population groups must take into account the possible contribution of several environmental agents as well as a background level of the disease.

Postulate 2: The organism must be isolated from the patients and grown in pure culture. It may be difficult or impossible to obtain the chemical in pure form from the patient. Chemicals often are metabolized and excreted in a different form. Adverse health effects may be the result of a combination of chemicals. Some of the chemicals may be unstable and difficult to isolate. The components of the mixture may vary over time, as is the case with polluted air or contaminated water. However, for a chemical such as lead, it is possible to find blood levels in exposed individuals and thus to correlate exposure with effect. (Because of genetic diversity, individuals do not respond exactly the same to the same stimulus, as we discussed previously, and so the correlation of exposure level with disease severity is not perfect. This is a fundamental reason why epidemiological studies require large numbers of people in order to determine cause-effect relationships.)

Postulate 3: The organism grown in pure culture must be able to cause the disease in a healthy host. In this context, this postulate says that the suspected chemical or chemicals in pure form must be able to cause the symptoms in appropriate animal (or other) models. The problem with this postulate is that causation of disease by chemicals is dose dependent. High doses often produce different symptoms and effects from low doses. Furthermore, measuring the effects of low doses of chemicals in a host animal requires a long time and a great many animals, whereas measuring the ability of an organism grown in pure culture to cause an infectious disease in a healthy host animal is a relatively straightforward process.

Postulate 4: The organism must be recovered again from the experimentally infected host. This postulate means that the mechanism of causation must be elucidated and some evidence of the suspected chemical (the chemical itself, a metabolic product, etc.) must be found in the affected subjects. Demonstration of the biochemical mechanism by which a chemical causes a disease is probably the most important step in proving a cause-effect relationship.

Despite all the problems and complexities involved in proving a true cause-effect relationship, it must be stressed that none of the four steps is essential for instituting preventive measures if there is reason to suspect that a chemical is doing human or environmental damage. As illustrated by Dr. Snow, practical preventive measures can be derived from strong association between environmental factors and incidence of disease. This is particularly true for chemicals that affect small numbers of people or take many years to show an effect.

STUDY DESIGN: PRECEPTS AND PITFALLS

Epidemiologists are guided by procedures based on sound principles developed through the many decades since the time of Dr. Snow and his colleagues. However, no cookbook sets forth the details of an epidemiologic study design. Every situation is different. The value of the results obtained from an epidemiologic study depends on the competence of the study design and the resources available for its proper conduct. Many factors must be considered in the design and conduct of a study. Examples of a few factors and the difficulties involved in the study of noninfectious agents are presented in the next paragraphs.

Perhaps one of the greatest difficulties in designing an epidemiologic study is recognizing and accounting for all of the possible factors that might be associated with the disease under investigation. This is a particular problem in modern society with diseases of unknown etiology. Populations are so diverse, particularly in the United States with its many cultures and ethnic groups, that many patterns of exposure can exist within a narrow geographical area. The epidemiologist must consider not only the more usual common exposures, such as air and water, but also occupational exposures, dietary habits, individualized drugs or medical treatments, hobbies and crafts, and recreational activities. If a critical factor is unknown or unsuspected, an epidemiologic investigation is unlikely to reveal its association with the disease.

Outbreaks of neonatal jaundice in the Imperial Valley of California in the early 1960s provide a good example of how a lack of complete information can interfere with the resolution of a problem. Neonatal jaundice is a disease that occurs among newborn infants as a result of a normal physiological process. The fetus has smaller amounts of oxygen available to it from the mother's blood than it will have from its own lungs after birth. The fetus compensates by making extra red blood cells to carry oxygen. After birth, it does not need as many red cells, so the excess cells are destroyed. The hemoglobin in the destroyed cells is degraded to bilirubin, a pigment that gives a yellow color to the skin, mucous membranes, and whites of the eyes. Destruction of old or excess red blood cells is a normal occurrence in humans, and adults have liver enzyme systems that rapidly destroy bilirubin and excrete it in the bile. These enzymes are either lacking or not present in sufficient quantities at birth. Perhaps as many as half of newborn infants develop mild, visible signs of jaundice.

In the early 1960s, many cases of serious neonatal jaundice occurred in babies born during the late summer months in a relatively new hospital in the Imperial Valley, surrounded by an agricultural area where a large amount of cotton was grown. Similar outbreaks had occurred during the same months for several prior years. Pesticides used extensively in the agricultural area surrounding the hospital came under suspicion as the cause of the outbreaks. Hospital records showed that the number of cases of neonatal jaundice increased among babies born at about the same time that cotton defoliants were sprayed on the surrounding fields. Thus, a possible cause was that the defoliant was causing liver damage in the newborns, which, in turn, was causing jaundice by preventing the breakdown of excess bilirubin.

Inspection of the hospital, which was new and modern, with few exterior windows, revealed no conditions within the hospital that prevailed only during the late summer months. Surveys of other hospitals in the same agricultural area showed no excess cases of neonatal jaundice, despite the facts that cotton defoliants and other agricultural chemicals were used to the same extent in the same months as in the vicinity of the affected hospital. Although the association with cotton defoliation was strong in the affected hospital, sound epidemiologic principles indicated that the cause of jaundice was not with agricultural chemicals.

An observant nurse in England independently noted that newborn babies in cribs near sunlit windows were less likely to suffer neonatal jaundice than infants in the interior of the nursery. Subsequent clinical investigation of the nurse's observation revealed that the degradation of bilirubin in newborns is hastened by exposure to bright daylight. Since the late 1960s, the routine treatment for neonatal jaundice has been exposure of affected babies to bright artificial light.

In retrospect, it seems so obvious that the additional cases of neonatal jaundice and the spraying of cotton defoliants in the Imperial Valley were both associated with the hottest summer months, when people of all ages stay indoors as much as possible to avoid sunlight. The fact that the hospital had fewer windows than other hospitals in the area was not deemed important at the time. However, the lack of windows combined with the propensity to keep newborns inside during the hot summer months reduced the babies' exposure to daylight to the point where they could not degrade their bilirubin at a rate sufficient to avoid jaundice. The clinical demonstration of the causal relationship between bright light and bilirubin destruction thus supports the theory that the Imperial Valley cases were due to a lack of bright light rather than to liver disease caused by exposure to cotton defoliants.

Another difficulty in designing epidemiologic studies of noninfectious agents is in grouping symptoms and counting the number of cases. It is important that apples not be counted as oranges, unless there is a valid reason to do so. How to group cases is a major problem in studying cancer clusters. Cancer is not a single disease; it is more than 100 different diseases with more than 100 different causes. In most cancer clusters, more than one type of cancer is usually represented. There may be cases of cancer of the brain, kidneys, bone, lymph system, and so forth. This poses a problem for biostatisticians; the smaller the number of cases of a single disease, the more data required to demonstrate an association with some environmental agent.

It is valid to consider different cancers as a single disease in order to increase the number of cases in a cancer cluster only if the environmental agent under investigation is capable of causing each of those different kinds of cancers, such as is the case with ionizing radiation. With few exceptions, carcinogenic chemicals are not random in their effects but rather cause specific cancers. Therefore, it is seldom valid to group different cancers when looking for an association with a chemical agent. Unfortunately, this difficult concept is one that news stories often ignore.

The collection of data is an especially important and sensitive aspect of a study design. Probably the least reliable method of obtaining information from people is by questionnaires that solicit information in the absence of an experienced interviewer. Such questionnaires ask about symptoms, history of illnesses, and history of exposures. Responses are difficult to evaluate because people differ in their ability to recall, they may not understand the questions, and their perceptions of the severity of symptoms may be tempered by their ability to tolerate illness. Thus, a valid epidemiologic study cannot be conducted from the comfort of an office. The people who design a study and collect and evaluate the data obtained should go into the field and talk to the people involved. They must explain the study and what its problems and potentials are. As in the case with the English nurse and the newborn infants with neonatal jaundice, people involved on the scene often have knowledge or make observations that are critical to the outcome of the investigation.

Information about the normal frequency of a disease in a population is essential before any determination can be made that its incidence is increasing or decreasing. Such information comes from statistics on morbidity and mortality, which often may not be as reliable as desired. If a disease does not normally occur in a population, the epidemiology of cases that do occur is simpler. For example, cholera has no normal background incidence in the United States because of good sanitation practices. An outbreak of cholera in this country would be quickly noted, and the cause would be relatively easy to find.

However, for noninfectious chronic diseases, such as cardiovascular diseases and cancer, there are background levels in all populations. The epidemiology of cancer and cardiovascular disease involves small differences in background incidence between populations or small changes within a population; hence the great importance of an understanding of what the background is. In recent decades, considerable data have been collected on age, sex, and geographic distributions of cancer and heart disease. These data provide the very valuable data bank required for epidemiologic studies of their association with environmental factors.

The importance of data on normal disease patterns to the interpretation of data obtained from epidemiologic studies is illustrated by studies of the relationship between exposure to pesticides and the incidence of cancer. It appears intuitively obvious that exposure to trace quantities of pesticides must be causing some disease. As a result, numerous studies have been conducted to compare cancer incidence between rural (presumed pesticide exposed) and urban (presumed not pesticide exposed) populations. A usual finding is a slightly greater incidence of certain leukemias and lymphomas among rural residents, despite the fact that overall cancer rates are higher in urban populations. A review of the literature is essential to determine the causes of differences in morbidity and mortality between rural and urban populations in order to evaluate these data.

Studies of vital statistics from years prior to 1960 and as early as 1943 in California, other parts of the United States, and other countries indicated that certain leukemias and lymphomas had a higher incidence in rural areas than in urban areas. Elfriede Faisal, a physician and medical epidemiologist with the California Department of Public Health, observed from these and her own studies that the differences were probably real and associated with exposure to the farm environment (*American Journal of Epidemiology* 87, no. 2 (1968): 267–274). The data reported in Dr. Faisal's paper were obtained in years prior to extensive use of synthetic pesticides in agriculture. In fact, many of the pesticides in current agricultural use were not commercially available prior to 1960. For proper interpretation, studies designed to investigate adverse effects of exposure to any agent must consider data from preexposure studies. To this date, the reason for the increase in some types of cancers in rural areas is still unknown. Further complicating the analysis, studies have shown that inner-city children may be exposed to high levels of pesticides due to spraying for roach and rodent control, which adds another layer of difficulty to the interpretation of rural versus urban pesticide studies (see P. J. Landrigan et al., "Pesticides and inner-city children," *Environmental Health Perspectives* 107, no. 3 (1999): 431–437).

As with any science, epidemiology can be misused, most frequently in efforts to support preconceived opinions about causal, or lack of causal, relationships. Even with the best of intentions, studies can be incomplete or faulty. It is very difficult for the average person to evaluate critically data obtained from epidemiologic investigations. Since prevention of disease is the ultimate goal of epidemiology, biased or incompetent studies are, at best, wasteful of resources and, at worst, detrimental to the public welfare. They do not provide information that can help in disease prevention, and they may needlessly mislead or inflame the public.

UNREASONABLE EXPECTATIONS

Public agencies employ epidemiologic techniques to investigate the relationship between the health status of a population and contaminants in air, food, and water. Such studies are a proper function of public agencies, and a great deal of excellent and valuable information can come from them if they are well designed, carefully planned, and conducted by professional epidemiologists. In fact, much of our knowledge of the human health effects of airborne contaminants in factory and urban air has come from epidemiology. Unfortunately, sometimes public agencies conduct quick-and-dirty studies that yield quick-and-dirty results, serving only to confound the issues rather than to provide answers.

The scenario in these cases often runs like this: The agency, usually at the behest of a legislative body, conducts a study in response to (or in anticipation of) a public outcry against the fouling of its air or water supply. The legislative body, not having an understanding of epidemiology, does not recognize the difficulties, complexities, and costs of such investigations. As a result, the request usually carries an unrealistic time limit of a few weeks or months and an inadequate budget. The public agency honors the request, usually because it is reluctant to tell the holder of the purse strings that it is ill-conceived. The result is often a meaningless study that is of no value because it neither finds statistically significant associations nor develops sufficient information to ease the worries of the community. Thus, a great deal of time and money are expended in producing data that do not help resolve the issue. The danger in such studies is that the public may be given a false sense of security or an unjustified anxiety and concern, depending on the relationship of the findings to the facts of the situation.

Even more likely and more damaging is the fact that inconclusive studies often lead people to lose confidence in the public agencies that serve them. Quite often, the public has already decided that a local chemical plant is the cause of cancer cases in the vicinity or that groundwater contamination is the cause of birth defects in the neighborhood before a public agency undertakes an epidemiologic investigation. The public, unable to understand why its officials cannot prove or disprove unequivocally the role of the chemical plant or the contaminated groundwater, decide that they are incompetent and that the original assessment of blame is accurate.

PROXIMATE EVENT APPROACH IN ASSIGNING CAUSE

Cause-effect relationships are often obscure and difficult to ascertain, even in the face of strong associations. People not trained in the analytic techniques of epidemiology often fail to understand that a valid association between a cause and its effect cannot be made from a single case (a one-person epidemic) unless there are other independent data that support it. As a result, the public tends to use the proximate event approach in assigning causes for effects. For example, if a person suddenly becomes ill after eating lunch or drinking a soft drink, it is human nature to blame the lunch or the drink, when the actual cause may very well have been something eaten the day or night before or exposure to the flu or some other infectious disease. Or a person may get the sniffles a few hours after visiting a friend at the hospital. The trip to the hospital (where everyone is sick) is blamed for the sniffles, when actually they are the result of a cold virus with an incubation period of 7 to 11 days.

Here is an example of proximate events leading to causation. A study published in 1998 by Wakefield et al. showed a correlation between vaccination and autism and used this correlation to conclude causation (A. Wakefield et al., "Ileal-lymphoid-nodular hyperplasia, non-specific colitis, and pervasive developmental disorder in children," *The Lancet* 35 (1998): 637–641). The investigation of the 12 children with the symptoms of autism in this study led to the conclusion that the measles/mumps/rubella vaccine caused autism as well as the various bowel diseases often associated with autism. After this paper was published, Wakefield made public statements to the effect that children should not be given these vaccines. An increase in measles in the United Kingdom and the United States resulted.

Ten of the paper's authors retracted the conclusions in the paper in 2004, and on February 2, 2010, *The Lancet* officially retracted the entire article (because of mistakes in interpretation and apparently unethical conduct on the part of Wakefield and two other authors, including being paid by parents who were suing the vaccine manufacturer, performing invasive tests on children without consent, and paying children to give blood). In May 2010, the United Kingdom's General Medical Council lifted the medical licenses of Wakefield and some of his coauthors.

Since Wakefield's original paper, the causation of autism by vaccines has been dispelled with many, many clinical trials. However, some parents still believe in this causation, probably because the obvious signs of autism become noticeable to parents at about the time the children are receiving their extensive series of childhood vaccinations. The vaccinations appear to be the proximal event, although earlier but harder-to-detect signs of autism already may have been present. As we mentioned earlier in this chapter, the mind is so disposed to equate correlation with causation that we willingly leap from the correlation to assuming that there is causation. No matter how many negative studies or retraction of studies are out there, some people will not be able to accept that the finding is coincidence.

The proximate event approach is occasionally correct. If a person drinks several glasses of fruit punch or some other acid drink that was made and stored in a galvanized container, the chances are excellent that he will become sick to his stomach very rapidly. The proximate event was the cause, and he would be correct in thinking so; sufficient zinc salts would have been dissolved from the galvanized coating by the acid liquid to make the punch emetic. However, the proof is not in the association between the punch and the illness but in the well-established fact that zinc salts are emetic and the demonstration of zinc salts in the punch.

Chemicals, particularly pesticides, may be innocent victims of the proximate event assumption. The cat next door gives birth to a malformed kitten. The day before, the lawn was sprayed with an herbicide; the herbicide is blamed for the malformation, when in actual fact the malformation must have occurred at some time earlier in the kitten's fetal development. However, the proximate event approach also may be correct for pesticides. A person who sprays insecticides over large areas without using protective equipment and observing proper precautions may well show symptoms caused by the chemicals used.

There is no solution to the problems that often result from attributing causes to proximate events. Misdiagnoses are so easy and apparently logical (and occasionally correct) that even physicians and scientists are sometimes guilty, particularly when the scientific data are sparse. Concluding causation is easier than searching further to determine if a proximate event is a probable or even a possible cause. Although little can be done to eliminate misuse of the proximate event approach, its existence and fallibility are worthy of a note of caution.

DISTRUST OF SCIENCE AND SCIENTISTS

Perhaps this is a good place as any to discuss some of the issues between the public and scientists and science in general. Public distrust is directed primarily against those sciences and attendant technologies that are viewed as sources of threat to human or environmental health and well-being. Many excellent, thoughtful essays on the subject of public doubts about science have appeared in scientific periodicals. The subject is very complex and of concern to many physical, biological, and social scientists. A great deal of profound thought on the part of some of the best minds in their respective fields has been devoted to the problem. Thus, it would be presumptuous to assume that the issue could be adequately discussed here. Rather, let us examine some of the probable reasons for public skepticism relating to issues within the field of toxicology in the hope that this will help replace distrust with open-mindedness, or at least with an attitude of tolerance to the benefit of everyone.

Reasons for public distrust or confusion may be the abundance of contradictory information presented in the news media on an almost daily basis and the media's seeming need for sensationalism. The conflict between the benefits derived from chemicals, drugs, foods, and vitamins and the hazards caused by their use will never cease to provide society and individuals with numerous dilemmas to resolve. Nitrites are necessary in cured meats to prevent the growth of the organisms responsible for deadly botulism, but epidemiologic evidence indicates that diets high in nitrites carry a greater risk of stomach cancer. The various birth control pill formulations have proven to be very effective in preventing unwanted pregnancy, but medical science warns of the risk of potentially serious adverse reactions with protracted use of some birth control pills. Women at the age of menopause who take estrogen compounds to relieve the distressing symptoms that occur with the change of life are warned that estrogen increases their risk of uterine cancer and heart disease, but estrogen may protect against osteoporosis in postmenopausal women. Examples of such dilemmas are endless. What information should a person seek?

The first question to ask is: Is there a substitute chemical or a different method for achieving the same end?" The next question is: What are the risks versus the benefits? Judicious questioning of medical societies, public health departments, federal agencies such as the Environmental Protection Agency (EPA) and the Food and Drug Administration (FDA), private agencies such as the American Cancer Society and the American Heart Association, educational institutions, libraries, the Internet, or any other appropriate individuals or groups will help provide the information necessary to evaluate the problem. However, it must be kept in mind that a few agencies, both public and private, actually have become advocates for one side or another of a particular issue despite their appearance of objectivity. Therefore, one should obtain as much information from as many different sources as possible.

It has become common in recent years for environmental or consumer groups to demand that chemicals they consider detrimental to their special interests be banned. Banning chemicals is a simplistic way of dealing with very complex problems and often produces worse problems than those sought to be remedied. In some cases, it leads to the elimination of a chemical whose toxic properties and hazards are quite well known and to the substitution of a chemical or chemicals about which there is much less information. The banning of chemicals denies our ingenuity to develop alternative methods of use that will be protective of the health of the public and of the environment, and it denies society the benefits such methods contribute. Proponents of bans either do not accept that the chemicals in question can be used safely or contend that the malevolence and greed of industry will not permit safe use.

Politics and special interest pressures often override and obscure the true nature of toxic hazards. In fact, no chemical is so toxic that it cannot be worked with safely, even if doing so requires the extremes of remote control operation or totally enclosed space suits with clean-air supplies. The aerospace industry has learned full well to work with some of the most toxic of materials. These toxic chemicals seldom are brought into ordinary channels of industry and commerce and so never come in contact with the general public. Like so many other matters relating to use of chemicals, the issue of chemical bans is one that affects the entire population and so should be settled by an informed society.

Fundamental to the question of why the public has come to distrust the sciences that deal with human and environmental health is the uncertain nature of those sciences. Unlike physics and mathematics, the health sciences cannot provide us with absolute or, in some instances, even reasonably certain answers. People do not question whether an apple will fall to the ground rather than fly skyward when it detaches from its tree. People do not question whether half a pie is less than its whole or that the whole pie is equal to all of the pieces cut from it. But people are skeptical about data concerning the health risks, or lack thereof, posed by exposure to synthetic chemicals. Why?

The question uppermost in the minds of individuals is whether exposure to some chemical or chemicals will be harmful to their health or that of their loved ones. Often this is the very questions that toxicology cannot answer with a definite yes or no. Science has no way of knowing the exact biochemical makeup of any individual person or precisely what quantity of chemical would be just below that person's threshold for the most subtle adverse effect of the chemical. Science can give an answer based on judgment, but it does not as yet (and may very well never) have the methodology to respond with direct experimental evidence in a human population. However, answers based on scientific judgment, no matter how well founded, often are rejected because people want absolute answers, not best guesses, to questions concerning their health. The issue is further confused because best guesses often vary among scientists. The validity of a guess depends on the expertise of the scientist who offers it.

The questions uppermost in the minds of legislators and regulatory officials relate to the nature and incidence of adverse effects that might result from exposure of large populations to very small quantities of environmental contaminants. Science does not have the resources—money, trained personnel, laboratory facilities, experimental animals—to provide such information for even a few, much less all, of the many chemicals we may encounter in our daily lives. Legislators and regulatory officials, like their constituents, want absolute answers, and they fail to understand why definitive experimental studies cannot be conducted to provide them. The next case, although old, is a typical example.

Let us consider the biological effects of low-level radiation insults to the environment, in particular the genetic effects of low levels of radiation on mice. Experiments performed at high radiation levels show that the dose required to double the spontaneous mutation rate in mice is 30 roentgens of X-rays. Thus, if the genetic response to X-radiation is linear, then a dose of 150 millirems would increase the spontaneous mutation rate in mice by 0.5 percent. Now, to determine at the 95 percent confidence level by direct experiment whether 150 millirems will increase the mutation rate by 0.5 percent requires about 8 billion mice. The number is so large that, as a practical matter, the question is unanswerable by direct scientific investigation (see Alvin A. Weinberg, "Science and trans-science," *Minerva* 10, no. 2 (1972): 209–222).

An effect that has a very low incidence of occurrence would almost certainly not be detected in an experiment using only a few animals. A study using 10 animals is insufficient to reveal an effect that occurs in only 1 out of 100,000 animals.

Skepticism is a natural consequence of the fact that the public expects more of scientists than they are able to provide. People not only want absolute answers, as described, but often answers that are in agreement with what they believe. For example, when a local dump or hazardous waste site is suspected of causing an excess number of birth defects, local officials or legislators often demand to know why the local health agency does not "just go in there and find out," not realizing the great difficulty of such a task. Or, when the public wants to know if local industrial air pollutants are causing lung cancer, it does not understand why an immediate answer cannot be given.

All too often in such cases, the public has already decided that the dump or waste site is causing birth defects or that industry effluents are causing lung cancer. Thus, if a biostatistical study of the pollution sources and rates of disease in the area indicates that there is no association between the pollution and the adverse effects, some people reject the report, claiming that a more competent study would show an association or that industry interests biased the results. Public skepticism often is intensified in such circumstances because of qualifications appended to study reports. Scientists who conduct well-controlled and well-conducted biostatistical studies usually state that their findings cannot be considered conclusive because of weaknesses and statistical uncertainty inherent in such studies. It is correct to point out these weaknesses, since even the best-designed studies, as we have mentioned, cannot always give a certain yes or no. In addition, public skepticism may be intensified because some industries do try to hide data that malign their products. The tobacco industry, which hid not only data about the effects of smoking but also, it appears, the toxicity of some of the compounds found in cigarettes, is an obvious example.

Another point to keep in mind is that no matter how large the experiment, if no effect is observed, one can say only that there is a certain probability that there is in fact no effect. One can never, with any finite experiment, prove that any environmental factor is totally harmless. This elementary point unfortunately has been lost in much of the public discussion of environmental hazards. This highlights one of the most frustrating matters with which toxicologists and health professionals must cope, namely, proving absolute safety of exposure to chemicals. It is most natural for people to demand assurance that the chemical exposures they experience are absolutely safe, and it is very difficult for them to grasp the fact that this is an assurance that no one is capable of giving. Absolute safety is the complete absence of harm. How does anyone prove the nonexistence of anything?

Negative toxicologic data always prompt further questions. If a toxicity experiment shows that a certain exposure is a level of no effect, the question can be asked: What if more animals had been used? If more animals yield the same result: What if the study had been continued through one or two generations? If a two-generation study yields the same results, then: What might happen after 5 generations? 10 generations? 20 generations? There is no such thing as a diminishing supply of questions; there is always one more unanswered one.

There is no such thing as a diminishing supply of skepticism either. A number of years ago, during a panel discussion on the use of herbicides in forestry, a man in the audience stated that the concept of the no-effect level was just not logical. He asserted that there must be a chemical somewhere in the world that would be harmful, no matter how small a dose was given. This observation seems to be rational. It is not unreasonable for people to believe that some chemicals may be harmful no matter how small the dose. However, there is no way to prove that such a chemical does or does not exist. For any chemical that does produce adverse effects down to the smallest dose that can be administered practically, there is no direct way to prove that some much smaller dose would be harmless or for that matter that it would be harmful.

The fact that toxicology cannot provide absolute answers to many of the questions that are of great concern to people should not be cause for alarm. Toxicologists can make judgments about the possibilities and probabilities of harm resulting from exposure to chemicals based on scientific data obtained from chronic toxicity testing, knowledge of the behavior of the chemical in animal systems, and application of appropriate margins of safety.

Scientists, like all human beings, can have widely differing political and social value systems. Some scientists find it difficult to separate their political and social attitudes, which they hold with great sincerity and conviction, from their science. Science is objective, but scientists are not necessarily so. The vehemence with which an otherwise rational scientist can hold and blatantly express a biased attitude was shockingly displayed many years ago at a meeting sponsored by the University of California to explore the subject of integrating chemical and biological pest-control methods in agricultural production. During his presentation, a noted population biologist stated that the petrochemical industry is at about the intellectual and moral level of the people who sell heroin to high school kids ("A spirited attack on pesticide use," *San Francisco Chronicle*, December 7, 1977, p. 5). Such inflammatory statements may be effective in gathering followers to a cause, but they are also destructive to the process of solving critical problems whose solutions are essential for the public good.

Social scientists have studied the role of bias in professional decisions, and numerous philosophical essays have been written on the subject. Some claim that it is not possible for scientists to divorce themselves from their social values when making scientific judgments. Yet the objectivity of scientists in their respective fields of scientific endeavor is of crucial importance to societal and political decisions that will be of benefit to society rather than to some ideological movement.

Perhaps bias engendered by political and social values is most prone to influence attitudes that relate to subjects outside the area of a scientist's expertise. It is hoped that most scientists can remain aloof from their personal biases when rendering judgments about the impact on society of scientific data from their own fields of endeavor.

Another and important (but often ignored and unstated) reason for disagreement about the potential hazards presented by chemicals is a result of the sources and supply of research monies. The federal government and other public agencies budget large sums of money to study the effects of chemicals on environmental and public health. Because funds from these sources are very limited, competition is fierce.

Money typically is granted for study in areas where the greatest problems are perceived. For example, a grant proposal to study the biochemistry of a chemical that does not appear to be toxic or carcinogenic probably would not compete successfully with a similar proposal for a chemical suspected of being a carcinogen, even though the former may have much more wide-ranging public health interest and significance. It is understandable that scientists whose support comes largely or totally from grants may stress the problems that their subject area poses for the environment or the public in order to secure funding for their studies.

The other large source of research funds comes from industry, which spends a great deal of money in developing the toxicity data required for chemicals that are regulated by governmental agencies. These monies may be spent on internally conducted studies or given to private toxicity testing laboratories or to universities. It is not in industry's self-interest, primarily from the point of view of product liability, to understate or hide the hazards posed by their products. Yet awarding grants or contracts to scientists who are openly antagonistic toward industry is also not in industry's best interest. It is difficult to balance these two objectives, both in selecting who will perform these studies and in trying to assess the validity of the data.

The government now requires that physicians publishing data or appearing before the FDA on behalf of a drug company publicize their financial connections to the industry in the hope that this will reveal bias. Many journals now require that grant money from whatever source be explicitly mentioned in the articles they publish on any scientific topic.

Health and other governmental agencies should bring the public into the evaluation and decision-making processes relating to health and environmental matters. People who have a sense of control over their own destinies are more likely to work cooperatively to solve problems and are more able to understand and appreciate the tremendous difficulties faced by science and by regulatory agencies in protecting environmental health. The public also has a responsibility to educate itself by studying all aspects of issues. People should investigate the backgrounds and fields of expertise of people who attempt to sway them to a point of view. Whether the view is accepted or rejected, a person who has studied all sides of an issue can feel confident that the course he or she chooses is the proper. An educated citizenry is the best insurance that decisions relating to environmental and public health protection will be of true and lasting benefit rather than just of immediate cosmetic value.

There are two rules for people who are interested in evaluating conflicting statements on adverse effects of chemicals:

1. *Know the source.* If statements are made by someone noted for espousing a particular cause, regardless of whether it is pro-chemical or anti-chemical in nature, beware: His or her bias may be coloring those statements.

2. *Become informed.* Learn all you can about the subject. Many good sources of information are available to the public. However, when searching on the Internet, be aware that the bias of the site may not be obvious. Just because you read something on the web does not mean it is true. Because easy access to many sites is so easy, it is difficult to sort through all the data. However, it is important to recognize true that an informed person is not at the mercy of propagandists and, as a result, is better able to make more effective and more competent decisions.

CHAPTER 11

THE STUDY OF RISK

Finally we come to one of the most difficult tasks that confronts toxicologists, physicians, and each individual: assessing risk. There is probably no human activity that does not carry with it some risk. However, it is only in recent years that risk has become something we frequently think about. What is the risk of having a heart attack? What is the risk of getting AIDS? What is the risk of an earthquake or a tornado? What is the risk of a nuclear disaster? What is the risk that greenhouse gases will cause the ice caps to melt? Our concerns run the gamut from personal to global. What do we mean by the word *risk*, and how do we evaluate it? And, when it comes to chemicals, not only what is the risk, but also what benefit does that chemical confer, and how do we as individuals and as members of society decide whether the benefit derived is greater than the risk taken? In this chapter, we define *risk* and look at some ways in which to understand the risk-benefit ratio.

Risk refers to the chance that some unpleasantness, injury, loss, or other harm will befall us or someone or something we are concerned about. A consideration of risk, either intentionally or unknowingly, probably has always been part of human thought processes. Even small children rapidly learn the risks of arousing parental ire. Hardly a day goes by that we do not consider the risk presented by some activity, even if it is as simple as not carrying an umbrella when the skies are threatening. Risk has many facets. We hear of risk, risk assessment, perceived risk, acceptable risk, unacceptable risk, risk-benefit, risk communication, risk management, and so forth. To understand these concepts better, we begin by looking at public health statistics, which provided the earliest systematic basis for evaluating risks that affected groups of people.

PUBLIC HEALTH STATISTICS

A major goal of public health agencies is the prevention of disease. In order to accomplish this, the nature of diseases and their incidence in the

The Dose Makes the Poison: A Plain Language Guide to Toxicology, Third Edition.
By Patricia Frank and M. Alice Ottoboni
© 2011 Patricia Frank and M. Alice Ottoboni. Published 2011 by John Wiley & Sons, Inc.

population must be known. The bookkeeping of public health serves this mundane but essential role by compiling vital statistics, the record of births, deaths, and morbidity. One use of vital statistics is to provide estimates of risk of occurrences of diseases.

The traditional method of expressing risk of disease has been based on historical information organized in the form of cause-specific death rates. These rates are calculated from the actual experience of a population group by counting the total number of people who died during a specified time period from a particular disease and dividing the deaths by the population at risk for that disease. The result, a number between zero and one, is the fraction of the population that died from the disease. To make this fraction a whole number, public health statisticians usually multiply the fraction by 100,000. The result is the number of deaths per 100,000 people in a specific population, say the United States or Europe. For example, the rate of death from breast cancer among Caucasian females living in the United States during 1990 to 1994 was approximately 0.00026, or 26 per 100,000.

Public health statistics can be used to compare the relative impact of various diseases, to examine the trend of diseases over a period of time, or to demonstrate the effectiveness of treatment of that disease. For example, in 1950, the death rate from Hodgkin's disease (lymphoma) among white males was approximately 2.3 per 100,000. Statistics for 1994 showed that the death rate decreased to 0.68 per 100,000, demonstrating the results of more effective cancer chemotherapy (see National Cancer Institute, Cancer Mortality Maps and Graphs, http://cancercontrolplanet.cancer.gov/atlas).

Such statistics also can be used to predict future risk presented by various causes of death. In making predictions, an assumption is made that next year will be about the same as past years or next year will follow the trend established by past years. Depending on the nature of the disease, such forecasts can be quite accurate. A good example of this is actuarial forecasting made by insurance companies. Such forecasts are based on statistical calculations of life expectancy and the causes and rates of death according to population age groups. Causes may be age, accidents, or other risks covered by the insurance company. Risks are calculated for the purpose of determining what premiums the company should charge for insurance. In making their estimates, actuaries consider factors that modify risk, such as age, smoking practices, obesity, and past history of disease. The financial strength of the insurance industry over a period of many years is evidence that risks for populations can be estimated with fair accuracy by using historical information on adverse outcomes and by continually adjusting these estimates as new information is available. Of course, the risk of any one individual dying during a particular time frame cannot be calculated with any certainty.

INHERENT RISK

The risk inherent in any situation or activity is one that cannot be modified by whether it is considered acceptable or unacceptable, how great or small it is perceived to be, or even whether it is recognized as a risk. Risk can be changed only by altering the conditions that produce the risk. For example, the risk of developing cardiovascular disease from eating a diet high in sugar cannot be changed by accepting or rejecting the risk, by considering it to be important or of no consequence, or by not knowing that it exists. The risk can be modified only by changing the diet or other factors responsible for it.

The actual chance that some harm or loss will occur, and how great or small the inherent risk is, cannot be known exactly, but in some cases, it can be approximated. For example, no one knows the true risk of any one 40-year old individual developing lung cancer from smoking for 20 years, but the chances can be estimated by examining the large body of data on the association between smoking and lung cancer. Risk estimates based on actuarial data can be quite accurate because they are based on records of adverse outcomes in large population groups. Further, the accuracy of the assumptions made for the estimates can be tested by how well they predict future occurrences.

However, many of the risks we face, particularly those associated with low-level exposure to environmental agents, cannot be estimated using traditional public health statistics. One reason is that morbidity and mortality statistics are based on medical diagnoses of causes of illness or death. The attending physician or autopsy pathologist can identify the immediate cause of death, such as bronchial cancer, leukemia, pancreatic cancer, or cardiovascular disease. At the present time, however, in many cases, medical science is unable to reveal what caused the cancer or other disease. This is especially true in the case of chemicals for which there are no known associations between exposure and some form of cancer.

Even with the very strong association between smoking and lung cancer, it can never be stated with certainty that a cancer in the lung of an individual smoker was caused by smoking. The reason is that although respiratory cancers, such as squamous cell cancer of the lung, are associated with smoking, they also occur in nonsmokers. If a squamous cell cancer occurs in a smoker, the chances are good that it was caused by smoking. However, it cannot be known for sure that the cancer would not have occurred if the person had not smoked. We all know this from the startling information that a friend who never smoked died of lung cancer while the fellow who smoked like a chimney is still going strong at age 85. If there were a true one-to-one correspondence between smoking and lung cancer *and there were no other causes of lung cancer*, then this could not have happened.

In the case of cigarette smoking, very large population studies have shown a clear association between smoking and increased rates of respiratory cancer. Studies with rodents have shown that smoke will produce higher rates of respiratory cancer. Some of the carcinogens in cigarette smoke have been identified and shown to cause cancer in test animals. Yet, as stated, despite all of this evidence, it is impossible to prove that a particular respiratory cancer in an individual was caused by smoking. It is possible only to say that, based on population statistics and animal experimentation, the risk of respiratory cancer in that person was very much higher than it would have been had he or she not been a smoker.

With the exception of some very rare cancers that appear to be caused only by specific carcinogens, it is impossible to determine the actual cause of any cancer in an individual. Similarly, birth defects can be recorded, associations determined, and risks of occurrence estimated, but the specific cause in individual cases usually cannot be determined.

There are no statistics for many of the risks we face, such as the risk of cancer from environmental exposures to chemicals, but there are no known cases of cancer (or birth defects) caused by exposure to trace quantities of any environmental chemical, natural or synthetic. As a general rule, medical diagnosis alone is unable to show a relationship between exposure to very small quantities of any chemical and cancer or birth defects. Thus, estimates of the risk of cancer or birth defects from exposure to trace amounts of a chemical cannot be based on a known incidence, nor can the accuracy or validity of the estimates be tested. In short, we can estimate the risk of occurrence of a particular kind of cancer, but we cannot accurately estimate the fraction, if any exists, of a particular cancer risk attributable to exposure to trace amounts of chemicals. The same is not true for certain medications (which are not found in the environment) for which animal studies clearly show that the drug produces a specific and rare defect and where the mother took the drug during the period of organogenesis (e.g., phocomelia produced by thalidomide).

It is important to note that a few diseases are caused by environmental agents that are directly diagnosable. Lead poisoning is a good example, even though the early symptoms of low-level exposure are vague and not particularly different from those of many minor illnesses. These symptoms include decreased physical fitness, tiredness, sleep disturbance, aching bones and muscles, abdominal pains, and constipation. Despite the vague symptoms, laboratory tests for levels of lead and certain biochemicals in blood can identify the cause of the illness and provide an estimate of the degree of poisoning.

Another example of an environmental agent that can be traced by the disease it produces is the organic chemical vinyl chloride. Studies of workers exposed to very low levels of vinyl chloride have shown that this chemical causes an extremely rare liver tumor known as angiosarcoma, which is almost never found in people with no exposure to vinyl chloride. Additionally, another relatively rare tumor, mesothelioma, usually can be attributed to asbestos.

RISK ASSESSMENT

When chronic toxicity testing became a common requirement less than a century ago, the emphasis was on what was a safe amount of exposure for the population rather than on the risks associated with trace quantities of chemicals in foods. Tests were designed to determine no-effect levels in animals, and permissible levels for human exposure were set by applying margins of safety to those levels determined by animal experimentation. Despite subsequent cries of outrage that so many carcinogens had been permitted to contaminate our food supply, the process worked well. We will use cancer as a model to explain the risk assessment process, but remember that this process works for other illnesses and for side effects from environmental chemicals and drugs.

During the past decade, a large number of chemicals has been retested using more sophisticated and more sensitive techniques. Except for a few chemicals that had a long history of use, such as some food color additives, retesting has not produced information requiring significant changes in regulatory procedures to protect public health. Despite the greatly increased use of synthetic chemicals in food production and processing during the past century, the general health of the nation has improved and life expectancy has gradually increased, increases that can be attributed to better nutrition, cleaner air, and better diagnosis and treatment of disease.

Increased public awareness that some chemicals can cause cancer created the demand for government regulation of chemical carcinogens and thus the need to estimate carcinogenic risk and to set safe levels of exposure to such chemicals. There are two very serious obstacles to such a task. One, mentioned previously, is the lack of data. Except for cancers resulting from occupational exposures and a few rare forms of cancer known to be caused only by specific carcinogens, there are no documented cases of human cancer from exposure to trace quantities of chemical carcinogens. Despite the conviction that trace amounts of chemicals, such as pesticide residues, are causing cancer, those cases cannot be identified because, even if they occur, they are so rare that they are hidden in the background incidence of cancer. If they cannot be identified, they cannot be counted.

The second obstacle to estimating carcinogenic risk and setting safe levels of exposure is the theory that there is *no* safe level of exposure and thus that regulatory agencies reject the use of classic no-effect levels and thresholds in establishing permissible levels of exposure. The decision that the initiation of a cancer process is a chance event requires a new approach for evaluating safety, a procedure called risk assessment.

Risk assessment is the procedure that is used to estimate the chance that an untoward event will occur. In cancer risk assessment, the untoward event is the occurrence of one or more cases of cancer resulting from exposure to a given quantity of carcinogen. Originally, the one-hit (chemical bullet) model was used to estimate such risk, as described in Chapter 7.

All available scientific data on the subject indicate that chemical carcinogens do not act on this simple chance basis. A number of other controlling or modifying factors must be included in carcinogenic risk analysis. This fact greatly complicates the risk assessment process. The simple one-hit model was expanded into a multistage model based on the premise that, in addition to one hit by a molecule of carcinogen, one or more other conditions must be present, simultaneously or sequentially, to initiate a cancer. The one-hit model might be likened to the proposition that an automobile will start when the ignition is turned on. A multistage model would add the requirement that the battery be charged and the tank contain fuel.

Numerous multistage statistical models have been developed to assess cancer risk from exposure to known or suspected carcinogens, because no one model seems to apply in all situations. To date, a few have been accepted as appropriate, but improvements and refinements continue. The multistage models currently used in cancer risk assessment assume that a chemical that causes cancer in an animal study can cause cancer in humans, that human exposure to the carcinogen is daily for a lifetime of 70 years, and that one molecule of carcinogen is capable of causing cancer. Based on these assumptions, assessments are then made to determine what level of carcinogenic risk is associated with various levels of human exposure to a specific chemical.

Those charged with protecting the public against dangerous exposures to carcinogens obviously must have some means for making estimates of what exposures are dangerous and what exposures would be virtually safe. However, in the process of estimating risk, mathematical models and formulas should be an adjunct to, not a substitute for, scientific judgment. In addition, regulatory agencies must be free from special-interest pressures if the standards they promulgate are to be based on concern for the public good, not political expediency.

A review of a large mouse study (see Chapter 7) by the Society of Toxicology (SOT) includes a concise, clear statement that addresses the issue of cancer risk assessment:

> The SOT Task Force recognizes the need for establishing acceptable exposure levels and the significant need for societal (regulatory) understanding of the effects of chronic low level exposure to potential carcinogenic materials. The Task Force further acknowledges that the determination of the acceptability of risk is a societal (regulatory) decision. However, the Task Force strongly asserts that the evaluation of toxicological responses (carcinogenic or otherwise) and the estimation of risk to a given exposure is a scientific endeavor and must be conducted free and separate from regulatory considerations. The scientific estimate of risk must use appropriate methods (models) to obtain "best" estimates for risk with levels of confidence clearly stated. The use of the most conservative methods (models) for estimating risk can be both scientifically inappropriate and misleading to those charged with the responsibility of setting levels of acceptable risk. When "best estimates" models are used which incorporate time to tumor data into risk assessment as opposed to only adjusting for time, the ED01 Study demonstrates that linear quantal models, i.e., non-threshold models, do not fit the data and non-linear models which often suggest practical thresholds provide a better expression of observed responses.
>
> —REEXAMINATION OF THE ED01 STUDY,
> *FUNDAMENTAL AND APPLIED TOXICOLOGY* 1, no. 1 (1981): 29.

The literature is replete with papers describing the various models used for carcinogenic risk assessment. The journal *Risk Analysis*, the official journal for the SOT, debuted in March 1981 in response to the increasing need for communication among scientists and regulators. This journal is an excellent source for anyone interested in more information on the subject of risk. It presents papers dealing with all aspects of risk, not just risk assessment procedures.

A good deal of knowledge of statistical methodology is necessary to understand the models used in risk assessment. People without such knowledge, but who want more information on risk assessment, will find the review written by Hopper and Oehme of the risk assessment process, its history, its definitions, and its methods a valuable introduction to the subject ("Chemical risk assessment," *Veterinary and Human Toxicology* 31, no. 6 (1989): 543–554).

Here is a real-life example of a cancer risk assessment performed by the California Office of Environmental Health Hazard Assessment. Acrylamide is a by-product that results when carbohydrates such as potatoes are cooked at high temperatures. The State of California has listed acrylamide as a cause of cancer and birth defects or other reproductive harm as required under Proposition 65, a California ballot initiative enacted in 1986 which requires the state to publish a list of chemicals known to cause cancer or birth defects. Let us assume for this example that the animal data on acrylamide are an accurate predictor of human toxicity.

Although it is possible to lower the amount of acrylamide in cooked foods, it is not possible to eliminate it altogether because of food processing methods. Unless you become a raw foodist, you will always ingest some acrylamide each day. Based on an analysis of fast-food french fries, various brands of potato chips, and other foods processed at high temperature, the risk assessment calculations found that if 100,000 people ate french fries once every 26 days for their entire lifetime, one additional case of cancer would result. For potato chips, one additional case would result if 100,000 people ate a serving of chips every 14 days. And for coffee, where acrylamide is produced in the roasting process, one cup every 3 days would lead to an additional case of cancer.

As you can see from this calculation, the effect of eating fried foods in moderation would be extremely difficult to detect, given the background incidence of cancer. However, if one takes into account the fact that many Americans eat fast food every day and that french fries are a large component of such meals, the increase in cancer due to acrylamide would be considerably larger. *But* the effect on obesity and the resulting decrease in life expectancy from eating fries daily is so much greater than the possible carcinogenic effect that when choosing your lunch or dinner menu, you should be more concerned about your weight due to the ingestion of all that fat and potatoes.

Before you run out and pitch all cooked carbohydrate-containing foods, remember that humans have been exposed to acrylamide ever since a person first tossed a potato into the fire and that the rat study showed a 10 percent increase in cancer at a dose about 900 times greater than the normal human diet. So, based on what we have discussed in this book concerning thresholds for cancer-causing agents, the applicability of animal data (especially rodent data) to humans, and perhaps even hormesis, it is up to you, the consumer, to decide whether to take the possible cancer risk by eating temperature-processed foods.

PERCEIVED RISK

How people perceive risk has been the subject of many social and psychological studies because of its importance in making decisions to control risks.

Legislators and regulators can obtain assessments of risk from scientists and statisticians. However, since the public often perceives a risk differently from the way it is described in risk assessment, legislators also must know how the public views the risk and the reasons for its perception.

Decisions by local government agencies are based on the public's perception of risk more often than on assessments made by scientists. A major goal of politicians is to be reelected, a goal that will not be achieved by antagonizing the electorate. Since local officials are closer to their constituents than state and federal officials, they are more vulnerable to public displeasure. For example, no matter how small a risk of reduced air quality is posed by a waste-to-energy conversion plant, and despite great need for management of household waste, some people will oppose having such a facility in their community because they consider the risk associated with incineration unacceptable, particularly in their own neighborhoods. The decision by local officials about building such a facility would depend largely on the strength of the opposition rather than on scientific or statistical risk assessment.

Risk perception is a very personal matter, and a very complex one. Just as no two people have the exact same fingerprints, so there are no two people who have the exact same perceptions of all possible risks. Risks may be viewed similarly by people who share the same value system, but many risks are not so linked. For example, some people are terrified of flying, but others enjoy it immensely without fear. Yet some people from both groups share a perception that smoking cigarettes is a very dangerous activity while others from both groups enjoy smoking and consider that the surgeon general's warnings are considerably overexaggerated or not pertinent to them. The risks associated with flying and with smoking, like so many other risks, are not coupled; they are viewed independently.

A brief mention of some of the factors that influence the perception of risk may be of value to understanding one's own perceptions, particularly those relating to risks associated with chemicals. The closer the public's perception of a specific risk agrees with the risk determined by an objective scientific and statistical risk assessment process, the greater the chance that regulatory decisions will be based on knowledge rather than on emotion, and the greater the chance that good and lasting benefit will accrue to society from the controls imposed. When all facets of society, industry, government, and the public agree on the magnitude of a risk, political decisions made to control the risk will be more generally acceptable, and cooperation with regulations will be easier to achieve.

Education is an obvious factor in risk perception. As a general rule, people who are highly educated in one discipline see the risks associated with the technologies developed from their disciplines much differently from people educated in other fields. Recent concerns about genetic engineering provide an example of how the difference in perception of risk depends on the education people possess. Microbiologists view the risk to public or environmental health from the manipulation of genetic composition of microorganisms differently from people educated in other disciplines. They know the processes, the significance of the alterations, the controls employed, and so forth.

People who have a college education perceive risks differently from people who do not. These differences cannot be attributed to less intelligence, since many people with above-average intelligence do not go beyond high school for a number of reasons, including financial resources, lack of interest in academic subjects, and family or peer pressures. The same can be said for blue-collar workers, who often see the risks associated with their occupations as being less than indicated by statistics on occupational illness and injury.

A primary source of information for a majority of people is the news: television, radio, print media, and the Internet. Television and radio commentators seldom have the luxury of delving into any news story with the breadth and depth required to provide viewers with enough facts to make informed decisions. Journalists who write for newspapers and magazines have more time to ascertain facts and more space to present a balanced story. Some may see themselves merely as reporters, but they are in fact educators. Those who write blogs may be biased and present information that supports their agenda.

Education is only one of many factors that influence perceptions of risk. Others can be of equal or greater importance, depending on the nature of the hazard and the risk involved. These factors include such diverse conditions as social status, economic status, age, sex, sexual orientation, national origin, place of residence, racial and cultural background, religious orientation, and the opinions of friends and relatives.

ACCEPTABLE RISK

The acceptability of a risk may be viewed on two levels, societal and personal. On the societal level, the process is generally something like this: The responsible governmental agency holds public hearings to obtain testimony from scientists and members of the public. After all information is gathered, the agency quantifies the risk using statistical methods described in the section on risk assessment. Next must come a policy decision as to how much risk is acceptable. In the case of cancer risk assessments, regulatory agencies are faced with the weighty responsibility of deciding how many cancer cases due to a carcinogen would be acceptable. Government officials, acting for society, have decided that one excess cancer in a population of a million constitutes an acceptable risk.

Several problems are associated with assigning a specific number of cases of cancer that are considered acceptable. One problem is the false impression that the figure is a matter of scientific fact rather than a statistically derived estimate. It delivers the erroneous message that one in a million people—no more, no less—will actually develop cancer from exposure to the chemical in question. It misleads the public into believing that one extra case of cancer in a population of a million people actually could be measured and the cause could be identified. Finally, it frightens people who fear that they or one of their loved ones may become that one unfortunate soul in a million. They wonder why their government would consider that any extra cases of cancer are acceptable. Stevie O. Daniels, former executive editor of the magazine *Organic Gardening*, articulated these concerns in her March 1989 editorial:

> The EPA defines negligible risk for adults as a one-in-a-million chance of getting cancer from a particular residue in a lifetime. ... That means roughly 231 people will develop cancer. ... What kind of a world do we have if we accept the incidence of cancer in one in every million people?

On a personal level, the determination that a risk is acceptable (or unacceptable) is quite a different matter. Often people reject societal decisions concerning acceptability. The organic food movement is a case in point. People who purchase only produce grown without the use of synthetic pesticides reject the societal decision that the risk of cancer from exposure to trace quantities of pesticides is negligible (acceptable).

As another example, statistical analysis shows definitively that wearing a helmet while on a motorcycle leads to reduced head injury in accidents, yet some people are willing to take on the risk of riding without a helmet, despite laws requiring the wearing of helmets in some states. Society is not willing for its people to take on this risk, but individuals may be. The risk of side effects from mosquito spraying may be quite small (and the benefit of preventing West Nile disease quite large), but because individuals do not have a choice in this matter—their streets will be sprayed whether they want them to be or not—they may be less willing to take on this small risk than the larger but personal risk of not wearing a motorcycle helmet or not buckling a safety belt. Personal decisions concerning the acceptability of risk are not based on statistical models but rather on how the risks are perceived. Whether the risks are familiar or foreign, voluntary or involuntary, result in a mild inconvenience or a disaster, and so forth, all play a role in their acceptance or rejection by individuals.

RISK BENEFIT AND COST BENEFIT

Risk benefit and cost benefit might be likened to two sides of a coin. The former balances the risk involved in some action or event against the benefit derived. The latter examines the cost of some action, such as reduction of a risk, in relation to the benefit achieved.

Almost every risk can be subjected to evaluation of benefits and costs from both personal and societal perspectives. Using the earlier example, the risk of not wearing a helmet when riding a motorcycle is suffering a massive head injury. For the individual, the benefit of not wearing a helmet might be the pleasure of feeling the wind blow through her hair. Some cyclists feel the benefit is worth the risk. For society, there is no benefit. Society has enacted laws requiring the wearing of helmets; such laws demonstrate its decision that, in the absence of benefit, the cost of caring for people permanently disabled with massive head injuries is unacceptable.

Chlorination of drinking water is another example in which there are individual and societal risk-benefit and cost-benefit considerations. Public drinking water supplies are chlorinated to kill pathogens that cause serious illness or death. The risk from chlorination is the production of trace amounts of chemicals, such as chloroform, suspected of causing cancer. The risk of not chlorinating public water supplies is the occurrence of mass outbreaks of waterborne diseases, such as cholera. For society, the former risk is negligible whereas the latter risk is unacceptable. The benefit of chlorination is considered to be greater than the cost. Most people probably give little thought to the matter. However, for a few individuals, the risk of exposure to trace quantities of carcinogens, no matter how slight, is unacceptable. For them, avoiding the perceived risk is worth the cost of bottled water. However, the risk from plastic eluting from these water bottles into the water should be accounted for in this personal equation, as well as the cost to society for disposal of the empty bottles. No risk-benefit equation is really as simple as we think.

The regulation of chemicals, particularly those that cause or are suspected of causing cancer, is the result of public demand for such controls. Few would dispute the need for government regulation of hazardous chemicals to protect public health from exposure to toxic and carcinogenic chemicals in workplaces, homes, and the general environment. In addition to the need for regulation, legislators recognize that risks, benefits, and costs must all be taken into account when setting standards for chemical exposures, because regulation of chemicals costs money—taxpayers' money. Unfortunately, people usually do not consider the cost of regulation and are not as involved as they should be in demanding rigorous risk-cost-benefit evaluations from their government officials.

The regulation of chemical carcinogens is a case in point. The cancer risk models used by regulatory agencies assume that one molecule of a carcinogen is capable of initiating a cancer process, that exposure to the carcinogen is a lifetime exposure (daily exposure for 70 years), and that data obtained from animal carcinogenicity studies are directly translatable to humans. Cancer risk estimates derived from these models are very conservative because of the nature of the assumptions on which they are based. It is usual and appropriate for public officials to err on the side of safety in matters relating to public and environmental health, if they are going to err at all. Regulatory agencies acknowledge that the cancer risk estimates currently being used for regulatory purposes have a large error on the side of safety. Thus, if the assumptions are incorrect, the errors can result only in excess safety for the public, not less safety. But if regulations are based on risk assessments that are in error by several orders of magnitude, the benefits to society could well be negligible or nonexistent and the costs to taxpayers could be overburdening.

As an example, the U.S. Department of Agriculture had a program in place for over 15 years that tested the concentrations of pesticides in various food crops; the data, along with Environmental Protection Agency (EPA) data on pesticide use, were collated and used to set acceptable food limits for pesticides. This program cost about $8 million per year. In 2008, the administration decided that budget cuts needed to be made and eliminated this program. The groups that used these data can now obtain the data from a private company for a fee, but many critics feel that the government data were more reliable and, of course, less expensive. Apparently, the government's risk assessment revealed that the cost of the analysis was not worth the benefit of knowing the pesticide levels.

Society has the right to decide on the level of safety it wants. It also has a right to know the cost increments for each increase in degree of safety so it can decide if the benefits are worth the cost. However, risk, benefit, and cost evaluations are proper only when the risk takers, benefit receivers, and pocketbooks are the same or equivalent. For example, assume that a person decides to paint his house himself to save money. One risk he faces is injury from falling off the ladder. The benefit is that he saves money. The risk and benefit are equitable—he takes the risk and he reaps the benefit. Now assume a factory decides to speed up an assembly line to increase profits. One risk is that a worker may suffer injury from the more rapidly running machinery. The benefit is increased profits for the company. In such a situation, the risk and benefit are not equitable—the worker takes the risk and the company receives the benefit.

The risk-benefit ratio for drugs is easier to calculate because drugs are taken for their benefit. Most of us don't really understand the risk involved and so we are unable to make a true risk-benefit calculation. For example, perhaps you have rheumatoid arthritis and are severely limited in your ability to perform the everyday tasks of life. The doctor recommends one of the newer tumor necrosis factor (TNF) antibodies on the market (e.g., Humira® or Embrel®). You read the label and see that a rare side effect of the drug is leukemia. Now you must make a risk-benefit assessment for yourself based on how rare is the leukemia, your age (so you can estimate for how long you will take the medication), your severity of impairment, the cost, and whether insurance will pay. In the end, most people with severe rheumatoid arthritis who can afford these expensive drugs end up taking them and benefit from a better quality of life. However, some people who are terribly afraid of getting cancer will refuse. Both choices are acceptable because each person has calculated his or her own level of risk and compared it to the expected benefit. What is not acceptable is to accept these drugs (or any drug) without examining the risk-benefit relationship.

But the choice is not always so easy. Take the case of anemia where the doctor recommends taking an iron supplement. Eating liver instead of taking the supplement might be an option. Liver is high in iron, is relatively inexpensive as a protein source, and no one ever overdosed on liver, Alternatively, iron supplement pills are inexpensive and are easy to take; but one can overdose on these pills, and iron overdosage is especially dangerous for children in two ways:

1. Children under the age of six can accumulate iron from too much daily supplement, which can lead to serious disease.
2. Iron pills, which look like little candies, may be stored where children can access them easily, leading to an acute overdose causing metabolic shutdown and leading to serious consequences.

It is estimated that over 20,000 children each year accidentally ingest iron, making iron overdose one of the leading causes of death by a toxic agent in young children.

RISK COMMUNICATION

Risk communication embraces the very delicate task of explaining all aspects of risks to the public. People are perfectly capable of understanding complex subjects that are of interest or importance to them if they are explained correctly. Unfortunately, government, industry, and academic scientists are not always capable of making such explanations understandable. Scientists and bureaucrats have much to learn about communicating with the public.

People also have a responsibility. They must appreciate that most scientists have little experience in communicating with the public and that some are intimidated by doing so. People should attempt to learn the meaning of some scientific words and terms to help them better evaluate conflicting opinions about risks. Some civic groups, such as the League of Women Voters, provide forums for interaction between scientists and the community because they recognize the need for public understanding of environmental issues. Industry, consumer groups, government, environmental groups, and civic groups must learn to communicate with each other to discuss openly and rationally the pros and cons of arguments relating to environmental matters.

Risk communication can carry with it its own special kind of risk for scientists who attempt to explain their science to the public objectively and in nontechnical terms. In doing so, they face the risk of being labeled apologists by one side or the other of an issue, and perhaps by both. This is particularly true in the fields of toxicology and carcinogenesis. Scientists who attempt to put the risks from exposure to trace quantities of synthetic chemicals into proper perspective are met with derision by environmental partisans. As a result, some scientists are hesitant to take part in public debates. Scientists face similar issues in the debate about evolution versus creationism or intelligent design, so many just stay quiet and out of the debate.

Journalists form one of the most important communication links between science and the public. Much of what the public knows about the risks and benefits of science and technology is obtained from television, newspaper, magazine, and Internet stories. Even scientists themselves gain some of their knowledge of other sciences from these same sources: the media. The importance of journalists in informing the public and, as a consequence, shaping public policy cannot be overstated. In *Health Risks and the Press* Victor Cohen, a senior writer for the *Washington Post*, has described the journalist's position:

> Whether we like it or not, we journalists have become gatekeepers. In some measure, our choices of what will be reported, and how the data will be reported, set the national agenda vis-a-vis health risks. In a sense, we have become part of the regulatory machinery. ... The very way we

report a situation can affect the outcome. If we ignore a bad situation or write a "no danger" piece, the public may suffer. If we write "danger" the public may quake.

—M. MOORE, ED., *HEALTH RISKS AND THE PRESS*
WASHINGTON, DC: MEDIA INSTITUTE, 1989.

Of course, this is also true of all other sources of news as well.

Journalists are often blamed for the public's misunderstanding, or lack of appreciation, of the safety or danger of some technology, when circumstances beyond their control may really be at fault. Journalists can communicate only that information given to them by their sources, who may not express themselves clearly, may exaggerate, or may voice conflicting opinions; experts of seemingly equal repute often disagree with each other. In addition, there are the ever-present time and space constraints that limit a journalist's ability to explain the nuances of a story. These constraints probably are also responsible for some instances in which scientists feel they have been misquoted or their words have been taken out of context. And finally, the public's penchant for stories that are brief, with dramatic or frightening headlines, guides editors, who are the final authority about what gets into print. We all know that the nine-second sound bite is what makes the news, regardless of how complex the story may be.

The majority of professional journalists are truly interested in serving their communities through the media. Those who attempt to explain both sides of issues so that readers can form their own opinions perform a valuable public service. Further, good journalists investigate the competence and biases of their sources so they can better evaluate the information they receive. Journalists who permit their personal biases to influence the content of their news stories can do a great deal of damage, particularly if they write well and interestingly. Unfortunately, some journalists who let their biases show in their reporting do not recognize their own failing. They may feel they are performing a public service, but if they misinform and mislead the public, deliberately or inadvertently, they are subverting an informed electorate. People who have access to reports of journalists who present all sides of an issue, particularly science issues, free of editorializing, are truly fortunate.

The popular press is also an important avenue for risk communication. A number of scientists, through lectures, essays, and books, have become as influential as journalists in forming public opinion. Some of these scientists describe the risks to society presented by science and technology in general and the petrochemical and drug industries in particular. Some have publicly aligned themselves with one or more environmental or social causes that they support with great fervor. Other scientists who are unaligned with any pro- or anti-chemical movement attempt to provide the public with a balanced perspective of the risks associated with chemicals in the environment. Several in the latter group attempt to counter the biases that feed poison paranoia by examining the arguments, actions, and motives of the anti-science and anti-technology movements.

Unfortunately, there are some people whose minds are closed to any idea contrary to their beliefs, regardless of whether they are convinced that some technology is perfect and has no drawbacks or that some technology is bad and has no virtues. There can be no communication with people with closed minds. Fortunately, they represent a minority of the public. The majority do have open minds and are willing to listen to opposing views. In scientific matters, as in every other facet of human endeavor and interaction, there is no substitute for communication.

RISK MANAGEMENT

Risk management is the control of risk by eliminating or modifying the conditions that produce it. People practice risk management in every aspect of daily life, often without realizing it. The parent who stores medicines and household chemicals out of a child's reach is practicing risk management. The driver who fastens a seat belt is practicing risk management. The hobbyist who provides good ventilation in the area where solvents are used is practicing risk management. The gardener who puts on protective clothing before spraying pesticides is practicing risk management. The chemist who puts on safety glasses before entering the laboratory is practicing risk management.

Governments practice risk management by passing rules and regulations that specify procedures for controlling risks and penalties for disregard of the procedures. The risks managed by governments are those that affect the public in general or specific groups of people. Businesses, industries, and public agencies become the risk managers by complying with the rules and regulations. However, before government can manage a risk, it must be recognized and described, and its importance to society must be evaluated. This process requires time, money, and considerable effort in laboratory and field research. The public, not recognizing the difficulties involved in the process, often becomes impatient with regulatory agencies for not acting rapidly enough to protect it or becomes angry at the costs involved. The risk management agencies that have suffered the greatest amount of criticism are EPA and FDA, both of whose jobs are to reduce risk to the U.S. population.

EPA was created in 1970 with the broad and almost unmanageable mandate to protect both public health and environmental health. It was created in response to public concerns that pesticides were destroying the habitat and the inhabitants of the planet. In its almost 40 years of existence, EPA has done fairly good job, despite the tremendous difficulties under which it has had to operate. It has restored the ecological balance in some locations that formerly had been sites of major water or air pollution and has prevented further deterioration in other areas.

The function of FDA was created in 1906 with the passage of the Pure Food Act, which contained regulations to prevent the sale of adulterated products. Consider that prior to this time, cocaine and arsenic, for example, were available over the counter in various foods and medicinal products. The passage of this seminal legislation and its various additional laws over the years has led to a considerably safer consumer environment in the United States compared with that found prior to this legislation and even today in many other countries. However, new problems arise every day that require FDA to be alert to issues such as the importation of adulterated food products (melamine in baby food and pet food from China) and medicinal products (new and serious side effects that require marketed prescription products to be removed from the market). *Regulatory Toxicology* (2001) edited by S. C. Gad details the whole regulatory history of FDA, EPA, the Consumer Product Safety Commission, and the Occupational Health and Safety Administration.

Risk management is most effective when combined with risk communication. Local industries that fail to communicate their activities to concerned citizens may face community antipathy or opposition, and perhaps even restrictive legislation by local governments. Companies that explain their problems and listen to the concerns of their neighbors with open two-way communication are those most likely to achieve successful risk management programs. Programs of risk communication and risk management that involve the community take considerable top management and staff time and can add to costs, but the benefits of community understanding and cooperation can more than offset the expenditure. Companies that are responsible citizens and the communities in which they reside both benefit from the relationship.

Now that you understand the science of toxicology and the methods used to gather toxicology data, you should be able to read those scary headlines with some degree of understanding and perform your own personal risk-benefit assessment. You can be an informed citizen, keep an open mind when you hear conflicting data and opinions, and remove some of the fear from your life. As we have seen throughout this book, chemicals are everywhere, and although they can be our friends, frequently their use can lead to health and environmental risks that we, as an informed public, need to assess in order to help our government and its agencies regulate rationally.

BIBLIOGRAPHY

BOOKS AND ARTICLES OF INTEREST

American Council on Science and Health, *PCBs: Is the Cure Worth the Cost?* revised May 1986.

Barrett, J. R. "Phthalates and Baby Boys: Potential Disruption of Human Genital Development," *Environmental Health Perspectives* 113, no. 8 (2005): A542.

Brisken, C. "Endocrine Disruptors and Breast Cancer," *CHIMIA International Journal of Chemistry* 62 (2008): 406–409.

Carroll, L. *Alice in Wonderland*, various editions.

Carson, R. *Silent Spring*, Matemia Riner, 2002 with forward by Linda Lear and afterword by E. O. Wilson.

Claus, G., et al. "Chemical Carcinogens in the Environment and in the Human Diet: Can a Threshold Be Established?" *Food and Cosmetics Toxicology* 12 (1974): 737.

Commoner, B. "The promise and perils of petrochemicals," *New York Times Magazine*, September 25, 1977, p. 38.

Fallik, D. "Lifesaving Tests You Can't Live Without," *AARP Magazine* (May/June 2009).

Federal Register 44, no. 52 (March 15, 1979): 15, 975.

Federal Register 44, no. 31 (May 31, 1979): 31, 514–31, 568.

Federal Register "Occupational Exposure to Formaldehyde: Final Rule," 52, no. 233 (December 4, 1987): 46, 205.

Freeze, R. A., and J. H. Lehr. *The Fluoride Wars: How a Modest Public Health Measure Became America's Longest-Running Political Melodrama.* Hoboken, NJ: John Wiley & Sons, 2009.

Gad, S. C., ed. *Regulatory Toxicology*, 2nd ed. New York: Taylor & Francis, 2001.

Hayes, A. W., ed. *Principles and Methods of Toxicology*, 4th ed. New York: Taylor & Francis, 2001.

Heiby, W. E. *Better Health and the Reverse Effect*, Deerfield, IL: MediScience, 1988.

Heiby, W. E. *The Reverse Effect.* Deerfield, IL: MediScience, 1988.

Hopper, L. D., and F. W. Oehme. "Chemical Risk Assessment: A Review," *Veterinary and Human Toxicology* 31, no. 6 (1989): 543–554.

The Dose Makes the Poison: A Plain Language Guide to Toxicology, Third Edition.
By Patricia Frank and M. Alice Ottoboni
© 2011 Patricia Frank and M. Alice Ottoboni. Published 2011 by John Wiley & Sons, Inc.

Janega, J. "Bird Haven Hides Traces of Poison," *The Chicago Tribune*, May 25, 2008.

Keilman, J. "Huffing Kills McHenry Teen," *The Chicago Tribune*, April 19, 2010.

Landrigan, P. J., et al. "Pesticides and Inner-City Children: Exposures, Risks, and Prevention," *Environmental Health Perspectives*, 107, no. 3 (1999): 431–437.

Linton, F. B. "Federal Food and Drug Leaders," *Food Drug Cosmetic Law Quarterly* 4 (1949): 451–470.

Mackie, D. et al. "No Evidence of Dioxin Cancer Threshold," *Environmental Health Perspectives* 111 (July 2003).

Maugh, T. H., II, "Research News: Cancer and the Environment: Higginson Speaks Out," *Science* 205 (1979): 51.

Moore, M., ed., *Health Risks and the Press*, Washington, DC: Media Institute, 1989.

NCTR Toxicology Study on AAF: *Journal of Environmental Pathology and Toxicology* 3, no. 3 (1980): 1–250.

Perrone, M., and C. Anderson, "Glaxo to Remove Zinc from Denture Cream," *AT&T News*, February 18, 2010.

Romano, J. A., et al. *Chemical Warfare Agents: Chemistry, Pharmacology, Toxicology, and Therapeutics*. CRC Press, Taylor & Francis Group, 2007.

Sathyanarayana, S., et al. "Baby Care Products: Possible Sources of Infant Phthalate Exposure," *Pediatrics* 121 (2008): e260–e268.

Schier, J. "Fatal Poisoning Among Young Children from Diethylene Glycol-Contaminated Acetaminophen–Nigeria, 2008–2009," *MMWR Weekly*, 58, no. 48 (2009): 1345–1347.

Schneider, M. *Journal of the American Dental Association* 89 (1974): 1092.

Sinclair, U. *The Jungle*. American Classics Series, 2008; originally published 1906.

Smyth, H. F. Jr. *Food and Cosmetics Toxicology* 5 (1967): 51.

Stolley, P. D., and T. Lasky. *Investigating Disease Patterns: The Science of Epidemiology Scientific*. American Library, 1995.

Stump, D. G., et al. "Developmental Neurotoxicicity Study of Dietary Bisphenol A in Sprague-Dawley Rats," *Toxicological Sciences* 115 (2010): 167–182.

Swan, S. H., et al. "Decrease in Anogenital Distance among Male Infants with Prenatal Phthalate Exposure," *Environmental Health Perspectives* 113, no. 8 (2005): 1056–1061.

Wakefield, A., et al. "Ileal-Lymphoid-Nodular Hyperplasia, Non-specific Colitis, and Pervasive Developmental Disorder in Children," *The Lancet* 351 (1998): 637–641.

Weinberg, A. M. "Science and Trans-Science," *Minerva* 10, no. 2 (1972): 209–222.

Weiss, B. "Neurotoxic Risks in the Workplace," *Applied Occupational and Environmental Hygiene* 5, no. 9 (1990): 587–594.

Williams, R. J. *Biochemical Individuality*. New York: McGraw-Hill, 1956, 1998.

WEB SITES OF INTEREST

Alternative Animal Testing
European Centre for the Validation of Alternative Methods (ECVAM): www.ecvam.
jrc.ec.europa.eu

Birth Defects
International Birth Defects Information Systems (IBIS): www.ibis-birthdefects.
org
Bureau of Alcohol, Tobacco, Firearms and Explosives (ATF): www.atf.gov

Cancer Information
American Cancer Society: www.cancer.org

Cancer Mortality Maps and Graphs
National Cancer Institute: http://cancercontrolplanet.cancer.gov/atlas

Chemistry, Molecular Structures
American Chemical Society: www.acs.org

Consumer Health Information
National Institutes of Health (NIH): www.nih.gov

Disease, Emergencies, Disasters
World Health Organization (WHO): www.who.int

Disease Prevalence and Prevention
Centers for Disease Control and Prevention (CDC): www.cdc.gov

Drug Clinical Trials
National Institutes of Health (NIH): www.clinicalresearch.nih.gov

Food Inspections
Department of Agriculture (USDA): www.usda.gov
Food and Drug Administration: www.fda.gov

Food, Nutrition, Chemicals
American Council on Science and Health (ACSH): www.acsh.org

Formaldehyde Information
Occupational Safety and Health Administration (OSHA): www.osha.gov/SLTC/
formaldehyde/index.html

Guidelines for Toxicology Testing
International Committee on Harmonisation: www.ich.org

Hazardous Material Transport
Department of Transportation (DOT): www.dot.gov
Also by each state

Human Genome
Human Genome Project: www.ornl.gov/hgmis

Malaria
Bill and Melinda Gates Foundation: www.gatesfoundation.org

Metric Conversion
www.metric-conversions.org

Morbidity and Mortality Reports
Center for Environmental Health at the Centers for Disease Control and Protection: www.cdc.gov/mmwr

Nutrition
Harvard School of Public Health: www.hsph.harvard.edu
Occupational Safety and Health Administration (OSHA): www.osha.gov

Organic Certification
National Organic Program: www.usda.gov or www.ams.usda.gov

Peanut Allergy
www.mayoclinic.com/health/peanut-allergy/DS00710

Periodic Table
Los Alamos National Laboratory: http.//periodic.lanl.gov
University of Sheffield: www.webelements.com

Pesticides
EPA fact sheet, "Protecting Children from Pesticides": www.epa.gov/opp00001/factsheets/kidpesticide.htm

Poison Control
American Association of Poison Control Centers: www.aapcc.org

Radioactivity Information
Idaho State University (http://www.physics.isu.edu/radinf/natural.htm)

Radon Information
Environmental Protection Agency: www.epa.gov/radon and www.epa.gov/radon/pubs/citguide.html

Space
National Aeronautics and Space Administration (NASA): www.nasa.gov

Work-Related Illness
National Institute for Occupational Safety and Health (NIOSH): www.cdc.gov/NIOSH

ABBREVIATIONS

AAF   acetyl amino fluorene

ACSH   American Council on Science and Health

ADI   acceptable daily intake

AHH   aryl hydrocarbon hydroxylase

ATF   Bureau of Alcohol, Tobacco, Firearms and Explosives

ALD   average lethal dose

C   Centigrade or Celsius

CDC   Centers for Disease Control and Prevention

DDD   dichloro diphenyl dichloroethane

DDE   dichloro diphenyl dichloroethylene

DDT   dichloro diphenyl trichloroethane

DF   dibenzofuran

DMSO   dimethyl sulfoxide

DNA   Deoxyribonucleic acid

DOL   Department of Labor

DOT   Department of Transportation

EPA   Environmental Protection Agency

F   Fahrenheit

FDA   Food and Drug Administration

FDCA   Food, Drug, and Cosmetic Act

FIFRA   Federal Insecticide, Fungicide, and Rodenticide Act

GRAS   Generally recognized as safe

HSLA   Hazardous Substances Labeling Act

ICH   International Committee on Harmonisation

IU   International Units

LC50   Lethal concentration for 50 percent of test animals

LD50   Lethal dose for 50 percent of test animals

The Dose Makes the Poison: A Plain Language Guide to Toxicology, Third Edition.
By Patricia Frank and M. Alice Ottoboni
© 2011 Patricia Frank and M. Alice Ottoboni. Published 2011 by John Wiley & Sons, Inc.

MAC	Maximum allowable concentration
MLD	Mean lethal dose
MSDS	Material Safety Data Sheet
NCI	National Cancer Institute
NCTR	National Center for Toxicological Research
NIH	National Institutes of Health
NIOSH	National Institute for Occupational Safety and Health
NOAEL	No adverse effect level
NOEL	No effect level
NOTEL	No toxic effect level
NRDC	Natural Resources Defense Council
OSHA	Occupational Safety and Health Administration
OTC	Over the counter
PBDE	Polybrominated diphenyl ethers
PCB	Polychlorinated biphenyl
PCC	Poison Control Center
RFD	Reference dose
RNA	Ribonucleic acid
2,4,5-T	2,4,5-trichlorophenoxy acetic acid
TCA	Tricarboxylic acid
TCDD	tetrachloro dibenzodioxin
TCDF	Tetrachloro dibenzofuran
TD1	Tumor dose for 1 in 100 individuals
TD50	Tumor dose for 50 in 100 individuals
TOCP	Tri-ortho-cresyl phosphate
USDA	U.S. Department of Agriculture
WHO	World Health Organization

GLOSSARY

abortifacient: An agent that produces an abortion.

abortus: The aborted products of conception.

adjuvant: A pharmacological agent added to a drug to increase its effect.

aerosol: A suspension of very small particles of a liquid or a solid in a gas.

alchemy: A primitive medieval science remembered primarily for its search for a method for converting base metals into gold.

amalgam: An alloy of mercury with another metal, the most commonly known of which is the silver amalgam used for dental fillings.

ambient: Surrounding. When applied to air, it means outdoor air.

amino acid: Organic compounds that contain an acid grouping and an amine grouping. Amino acids are the building blocks of proteins.

analogous: Similar or resembling each other in some way.

anatomic: Pertaining to the structure of an organism.

angina: Pain in the chest; usually taken to mean heart pains.

anomaly: A thing or organism that deviates from normal.

antagonism: An interaction between two chemicals that results in one lessening the toxic effect of the other.

anthropomorphic: Attributing human characteristics or form to nonhuman things.

anticoagulant: A chemical that prevents clotting of the blood.

aqueous: Watery, or pertaining to a water solution.

aromatic: In organic chemistry, compounds that contain one or more benzene rings.

aspirate: To inhale liquid into the lungs.

atom: The smallest unit of an element that still maintains the physical and chemical properties of the element.

atrophy: To decrease in size or waste away.

The Dose Makes the Poison: A Plain Language Guide to Toxicology, Third Edition.
By Patricia Frank and M. Alice Ottoboni
© 2011 Patricia Frank and M. Alice Ottoboni. Published 2011 by John Wiley & Sons, Inc.

benign: Harmless.

bile: A digestive liquid secreted by the liver.

bile duct: The tube that carries bile from the liver to the small intestine.

biochemical: A chemical produced by a living process.

biodegradable: Capable of being metabolized by a biologic process or organism.

biodegradation: The process whereby chemicals are broken down by a biologic process or organism.

caffeine: A naturally occurring stimulant found in coffee, tea, and cola nuts.

carcinogenic: Capable of causing cancer.

cardiovascular: Pertaining to the heart and blood vessels.

castration: Removal of the testes or ovaries or their function (chemical castration).

chloracne: A skin disease, resembling the acne of adolescence, caused by exposure to chlorinated aromatic organic compounds.

chlorinated: Pertaining to the presence of one or more atoms of the element chlorine in a compound.

chromosome: One of the group of structures that form in the nucleus of a cell during cell division. Chromosomes, composed of DNA, carry the genetic code for the organism.

-cide: A suffix meaning "killer."

cilium, cilia (pl.): Microscopic hairlike cellular projections that are capable of sweeping movement.

colic: Acute abdominal pain.

colitis: Inflammation of the large intestine (colon).

compound: A chemical substance composed of molecules all of the same kind.

confidence level: A statistical term that expresses how assured one can be that results obtained from an experiment did not occur by chance. For example, a 95 percent confidence level says that there is a 95 percent probability that the results obtained were due to the conditions of the experiment and a 5 percent probability that they were due to chance.

congenital: Pertaining to a condition existing before or at birth.

coumarin: A naturally occurring anticoagulant compound.

cytoplasm: Cellular material within the cell membrane and surrounding the nucleus.

DDD: Dichloro diphenyl dichloroethane, a metabolite of DDT and itself a pesticide.

DOE: Dichloro diphenyl dichloroethylene, a metabolite of DDT.

demyelinate: To destroy or remove the sheath of fatlike material (myelin) that surrounds certain types of nerve fibers.

dermatitis: A skin inflammation.

dichotomy: A division into two mutually exclusive groups.

dielectric: Pertaining to the nonconductance of electricity.

differentiation: In biology, the process whereby cells develop individual characteristics and become specialized in form or function, that is, liver cells, kidney cells, muscle cells.

dioxin: A group of chlorinated organic compounds formed during combustion or as a by-product in the manufacture of various chlorinated products.

DNA: The biochemical molecules (deoxyribonucleic acid) from which chromosomes are made; chromosomes, located in cell nuclei, carry the genetic code.

dosage: The regulation of doses, that is, how often, for how long.

dose: The quantity of chemical administered at one time.

dose-response curve: A graphic representation of the relationship between the dose administered and the effect produced.

dust: Very small solid particles generated by grinding, crushing, or other mechanical processes.

edema: Swelling due to collection of fluid in tissues.

electrocardiogram (ECG): The recording of heart electric function (occasionally abbreviated as EKG from the original German).

electroencephalogram (EEG): The diagnostic recording of brain waves.

electron: A subatomic particle that carries a negative charge.

element: A chemical substance composed of atoms of all the same kind.

embryo: An organism in the early stages of its development. In the human, it is the developing individual from conception to the end of the second month of uterine life.

emetic: An agent that induces vomiting.

empirical: Based on experience and observation.

endemic: Pertaining to prevalence in a particular region. An endemic disease is one that is constantly present at a specific incidence in a particular region.

endocrine disruptor: An exogenous compound that can act like a hormone in the body, leading to interference with the body's own hormone function.

enzyme: A biochemical, usually protein, that speeds up the rate of biochemical reaction.

epidemiology: Originally, the science that studied the cause and control of epidemics, outbreaks of a communicable disease in a region. Now its subject matter includes diseases caused by chemicals and other environmental factors.

epigenetic: A change in gene expression that does not involve a change in DNA.

esophagus: The tube that connects the mouth with the stomach.

estrogens: A group of female hormones.

ethanol: A naturally occurring 2-carbon alcohol, usually obtained from a fermentation process, and commonly known as alcohol.

etiology: The study of the cause of a disease.

extrapolation: The process of estimating unknown values from known values.

extrauterine: Outside of the uterus.

feral: Undomesticated; living in a wild state.

fetotoxic: Toxic to a fetus.

fetus: The later stages of a developing organism. In the human, it is the unborn child during the period of uterine life from the end of the second month until birth.

fume: Very small solid particles generated by recondensation of a vaporized solid.

fungicide: An agent that kills fungi such as molds, mildews, and mushrooms.

gas: Individual molecules or atoms of a substance that has a boiling point below normal room temperature.

gastric: Pertaining to the stomach.

gastrointestinal: Pertaining to the stomach and intestines.

gene: The smallest subunit of a chromosome that contains a genetic message.

genome: The complete set of hereditary factors of an individual.

genotoxic: Damaging to genetic material.

germicide: An agent that kills germs (pathogenic microorganisms).

germinal: Pertaining to the reproductive cells.

gestation: Pregnancy.

gonad: An organ that produces reproductive cells.

half-life: The length of time required for the quantity of the matter or property in question to be reduced by half.

hemoglobin: The red-colored biochemical in red cells that transports oxygen from the lungs to the tissues.

hemolysis: The rupture of red blood cell membranes that permits their contents to escape into the surrounding fluid.

hepatic portal vein: The blood vessel that carries blood from the intestinal walls to the liver.

hepatocarcinoma: Cancer of the liver.

herbicide: An agent that kills plant life.

histopathology: The study of microscopic abnormalities produced by diseases.

homologous: Structures that are similar in position or function.

hormesis: A positive response to low levels of a toxicant.

hormone: A biochemical secreted by one tissue in the body that exerts an influence on a biochemical function or organ somewhere else in the body.

hydroxylate: To introduce an oxygen-hydrogen group (hydroxyl group) into a molecule.

hyperpigmentation: The condition of having excess color in tissues or organs, such as the skin.

hypertrophy: Excessive enlargement or overgrowth of an organ or tissue.

hypervitaminosis: An abnormal condition due to the excessive intake of one or more vitamins.

insecticide: An agent that kills insects.

isomer: A molecule that has the same number and kind of atoms as another molecule but has a different arrangement of the atoms.

isotopes: Atoms of the same element that differ in weight; for example ^{11}C and ^{12}C, both carbon, but ^{11}C is radioactive.

in silico: Experimentation via the computer.

in utero: In the uterus (womb).

in vitro: In glass (test tube) or otherwise outside of a living organism.

in vivo: In a living organism.

lipids: Organic chemicals that possess certain properties, such as being insoluble in water, commonly known as fats.

lymphatic system: A system of vessels that originate in the tissues, drain the tissues of their clear fluids (lymph), and return the fluids back to the bloodstream at a site near the heart.

macromolecule: An extremely large molecule.

malaise: A vague feeling of discomfort or debility.

malignant: Very injurious or deadly.

mass spectrometry: A sensitive analytical method for determining the concentration of a chemical compound.

menopause: Literally, the cessation of menstrual periods. Menopause signals the end of a female's fertile life. In the human, this period is also referred to as the change of life.

mesothelioma: A form of cancer caused by exposure to asbestos, commonly found in the lining of the pleural cavity.

metabolism: The sum total of the biochemical reactions that a chemical undergoes in an organism.

metabolize: To undergo biochemical change in an organism.

metastasis: The spread of a disease to another part of the body.

methemoglobin: An oxidized form of hemoglobin that is not capable of transporting oxygen.

millirem: A thousandth of a REM (roentgen equivalent in man), a unit of ionizing radiation that, when delivered to humans, is biologically equivalent to one thousandth of a roentgen of X- or gamma radiation (see roentgen).

mole: The molecular weight of a compound expressed in grams.

molecular weight: The weight of a molecule, calculated by adding the individual weights of all of the component atoms.

molecule: The smallest unit of a compound that still retains the physical and chemical properties of the compound.

monosaccharide: The simplest molecules of sugars are the building blocks of complex sugars, such as starch and glycogen.

morbidity: The relative incidence of a disease in a population.

mortality: The relative incidence of deaths in a population.

mutagenic: Pertaining to the ability to produce change, particularly genetic change. A compound that is mutagenic is referred to as a mutagen.

mutant: An organism that has undergone a genetic change.

narcosis: The state of stupor or unconsciousness produced by a chemical.

nasopharynx: The area in back of the nose and above the opening of the mouth into the throat.

nausea: Stomach upset accompanied by a feeling that one is about to vomit.

neoplasm: Literally, a new growth. A neoplasm, also called a tumor, results from a more rapid than normal division of one or a few cells and may be benign or malignant.

no-effect level: As used in this book, a quantity of chemical that is below the threshold on the dose-response curve. In toxicology, the NOAEL refers to the no observed adverse event level while the NOTEL refers to the no observed toxic event level.

ocular: Pertaining to the eye.

oncogene: A gene that, when mutated or expressed at high levels, helps turn a normal cell into a tumor cell.

origin: The point on a graph that represents zero on both the vertical and horizontal axes (lines).

osmolality: When the concentration of ions in two solutions are the same.

osteoporosis: A condition in which bones become less dense, and hence more fragile.

oxalic acid: A naturally occurring organic acid.

ozone: A gas composed of molecules of triatomic oxygen (0^3), the most reactive form of oxygen.

pandemic: Epidemic disease spread over many regions, such as H1N1 flu during 2008–2009.

paranoia: A mental disorder characterized by delusions.

parasympathetic: Pertaining to part of the nervous system below the level of consciousness that participates in the regulation of the involuntary functions of the body; for example, heartbeat, breathing rate.

pathologic: Pertaining to a disease state. Pathology is the study of disease.

pentose sugar: A monosaccharide containing 5 carbon atoms in its structure.

peripheral nerves: The nerves that occur in the outer parts of the body, as opposed to those that occur in the brain and spinal column.

peritoneal cavity: The abdominal cavity that contains such organs as stomach, intestines, and liver.

pesticide: An agent used to kill pests. The category pesticide contains many -cides.

petrochemical: A chemical derived from petroleum.

pH: A term used to express the degree of acidity or alkalinity of a solution. A pH of 7 is neutral. Acid solutions have a pH below 7 and alkaline solutions have a pH greater than 7.

pharmaceutical: A therapeutic drug.

phylogenetic: Pertaining to the evolutionary relationship among organisms.

physiologic: Pertaining to the functions performed within living organisms.

placenta: The organ that forms the bridge between the fetal and the maternal bloodstreams.

pneumonitis: Inflammation of the lungs.

poison: A chemical that is acutely highly toxic.

polycyclic: In chemistry, pertaining to organic compounds whose molecules contain more than one circular structure (ring).

polymer: A large molecule (macromolecule) formed from the chemical combination of many smaller molecules, usually of only a few kinds. Proteins, starch, and cellulose are examples of natural polymers.

polyneuritis: Inflammation of many nerves at the same time.

potable: Suitable for drinking.

precursor: In biochemistry, a chemical from which another chemical is made.

primate: A member of the primate order of mammals, including humans, apes, and monkeys. *Nonhuman primate* refers to all the primates except humans.

prognosis: A forecast as to the probable outcome of an illness.

protocol: In science, the rules and outline of an experiment.

puberty: The period in which sexual maturity is attained.

purine: A family of complex organic chemicals composed of carbon, hydrogen, and nitrogen, arranged in two rings.

pyrimadine: Like the purines, except that the structure contains only one ring.

qualitative: Pertaining to kind or type.

quantitative: Pertaining to amount or degree.

quicksilver: The ancient name for the element mercury.

radiomimetic: Imitating ionizing radiations.

rectum: The portion of the large intestine (colon) just before it opens to the outside of the body through the anus.

RNA: The biochemical molecules (ribonucleic acids) present in the cytoplasm of cells that function with DNA in carrying genetic information.

rodent: One of a group of gnawing mammals that includes rats, mice, guinea pigs, and hamsters. Rabbits used to be considered rodents but are now in the group known as lagomorphs.

rodenticide: Aan agent that kills rodents.

roentgen: A unit of X- or gamma radiation that describes the degree of ionization that results under certain specified conditions, named for the discoverer of X-rays, Wilhelm Roentgen (1845–1923).

respirable: Capable of being inhaled.

sassolite: A mineral form of the element boron.

scrotal: Pertaining to the anatomic sac that holds the testes.

solanine: A naturally occurring chemical that occurs in potatoes and related species.

somatic: Pertaining to body cells as opposed to reproductive cells.

substrate: A chemical that serves as the substance acted upon by an enzyme.

synergism: An interaction between two chemicals that results in one enhancing the toxic effects of the other (scientific definition).

synthetic: Made by humans.

systemic: Pertaining to parts of an organism that unite in a common function.

teratogenic: Pertaining to the ability to produce birth defects.

teratologist: A scientist who studies the causes of birth defects.

teratology: The study of abnormal embryological development and congenital malformations.

therapeutic: Pertaining to the art of healing.

therapeutic index: The toxic dose of a drug divided by its curative dose. The greater the ratio, the safer the drug.

threshold: As used in this book, the point on a dose-response curve above which effects occur and below which no effect occurs.

tolerance: The concentration of a pesticide residue or food additive permitted by regulations of the Environmental Protection Agency and the Food and Drug Administration, respectively, to be in a specific food product.

uterine: Pertaining to the uterus (womb).

vector: In biology, an object, organism, or thing that transmits disease from one host to another.

volatile: Readily convertible to a vapor or gas form from a liquid or solid form.

xenobiotic: Foreign to life.

APPENDIX A

In order to understand how to calculate the number of molecules in any quantity of a chemical, two terms must be defined: mole and Avogadro's number.

A *mole* of any compound is its molecular weight expressed in grams. The molecular weight of a compound is the sum of the weights of its component atoms. For example, a water molecule is composed of 2 atoms of hydrogen (atomic weight of hydrogen is 1) and 1 atom of oxygen (atomic weight of oxygen is 16); thus, the molecular weight of water is 18, and 1 mole of water weighs 18 grams, or a little over 0.5 ounce. A molecule of table salt is made up of 1 atom of sodium (atomic weight of sodium is 22) and 1 atom of chlorine (atomic weight of chlorine is 35); thus, 1 mole of salt weighs 57 grams, or a little less than 2 ounces.

Avogadro's number is a confusing concept because it is so very large: 6.02×1023. (This scientific notation means 6 with 23 zeros after it; conversely, a number with superscript −23 as 6×10^{-23} means 23 zeros between the decimal point and the 6—a very small number indeed). But thinking of Avogadro's number as a term that refers only to a specific number, like dozen or gross makes it easier to deal with. A dozen means 12, a gross means 144, and Avogadro's number means 6×1023.

The number of molecules in a mole of any compound is equal to Avogadro's number. The weight of a mole will vary depending on the compound, just as the weight of a dozen will vary depending on what the dozen consists of, but the number of molecules in a mole is always the same. A dozen automobiles weigh tremendously more than a dozen eggs, but there is the same number of each. Thus, there are 6×1023 molecules of water in 18 grams of water and 6×1023 molecules of salt in 57 grams of table salt, even though the weight of 1 mole of water is only about a third of the weight of 1 mole of salt. By the way, October 23 (10/23) is National Mole Day, which is observed from 6:02 A.M. to 6:02 P.M. with a variety of celebrations worldwide.

The Dose Makes the Poison: A Plain Language Guide to Toxicology, Third Edition.
By Patricia Frank and M. Alice Ottoboni
© 2011 Patricia Frank and M. Alice Ottoboni. Published 2011 by John Wiley & Sons, Inc.

245

A charcoal-broiled steak contains about 10 μg of benzo[a]pyrene. The number of molecules in 10 μg benzo[a]pyrene is calculated in this way: The molecular weight of benzo[a]pyrene is 252; therefore, 1 mole would weigh 252 grams; 252 g is equal to 2.52×10^8 μg. The number of molecules in 1 μg is obtained by dividing the number of molecules in 1 mole by the number of μg in a mole. For benzo[a]pyrene, 6×10^{23} molecules in 1 mole divided by 2.52×10^8 μg in 1 mole equals 2.4×10^{15} molecules in a μg. The number of molecules in 10 μg benzo[a]pyrene would be 10 times as much, or 2.4×10^{16}.

INDEX

The Dose Makes the Poison: A Plain Language Guide to Toxicology, Third Edition.
By Patricia Frank and M. Alice Ottoboni
© 2011 Patricia Frank and M. Alice Ottoboni. Published 2011 by John Wiley & Sons, Inc.

Labels, labeling. *See also* Hazardous
   Substances Labeling Act; MSDS
   consumer products, 20, 50
   dangerous substances, 21, 37–39, 49,
      110–111, 145
   drug, 6, 26, 28, 124
   organic, 42–43
   poison, 7, 109
Laboratory, laboratories
   analytical, 42, 96–99, 176
   chemical, 4–7, 12, 16, 174
   research, 1, 3, 20, 68, 82, 163, 188, 206,
      209, 227
   toxicological, 40, 48, 92, 93, 104, 106,
      181
Laboratory Animal Welfare Act, 99
Laws, *see specific laws;* Legislation
LC$_{50}$, LD$_{50}$
   body weight and, 105
   corrosiveness and
   for DDT, 71
   defined, 48, 85, 103–104
   or DFs, 170
   humans and, 105–106, 116
   method for, 83–86, 107–108
   for parathion, 49
   for sodium chloride (salt), 50
   for sodium fluoride, 49
   for TCDD, 166
   for vitamin D, 49
Lead, xi, 32, 48, 98, 113, 122, 126, 159,
      161, 177, 178–179, 186, 196, 214
Lead arsenate, 36
Legionnaires' disease, 182
Legislation
   Best Pharmaceuticals for Children
      Act, 72
   Emergency Planning and Community
      Right-to-Know Act, 39
   Federal Caustic Poison Act, 37
   Federal Insecticide Act, 36
   Federal Insecticide, Fungicide, and
      Rodenticide Act (FIFRA), 37
   Food Additives Amendment Act
      (Delaney Clause), 69
   Food, Drug, and Cosmetics Act
      (FDCA), 37
   Food Quality Protection Act, xvi, 69

Good Laboratory Practices Act, 40,
Hazard Communication Standard, 39
Hazardous Substance Labeling Act
   (HSLA), 38, 50
Laboratory Animal Welfare Act, 99
Miller Amendment to FDCA, 37
Occupational Safety and Health Act
   (OSHA), 39
Poison Prevention Act, 38
Pure Food and Drug Act, 36–37, 228
Superfund Amendments and
   Reauthorization Act (SARA), 43
Lethal concentration, *see* LC$_{50}$
Lethal dose, *see* LD$_{50}$
Light, sun or ultraviolet, xv, 19, 26, 33,
   66, 74, 79, 110, 133–134, 141, 165,
   198–199
Linton, F. B., 36
Liver
   damage or disease of, 48, 67, 70, 71,
      73, 116, 166, 171, 173, 178. *See also*
      specific chemicals
   enzymes, 13, 63–65, 71, 116, 140, 172,
      198–199
   as a metabolic organ, 27, 58, 63
   neonatal jaundice, 198
   tumors and/or cancer, 145, 170, 172,
      176–177, 215
Liver cancer, *see* Liver tumors
Los Angeles Medical Association Poison
   Control Center, 109
Lung cancer
   air pollution and, 207
   asbestos and, 77
   from radon, 23, 134
   rates, 143, 193
   smoking, tobacco use and, 75, 147,
      151, 213–214
Lungs
   absorption by, 54, 56–57, 60, 62–63,
      114, 137
   aspiration into, 52
   damage or disease of, 23, 26, 55, 73,
      147, 152, 197
   excretion by, 65, 184
   granulomas, 57
   petroleum distillates and, 27
   surface area of, 55

TCDF, 170, 171
Teratogenesis, teratogenicity, 118, 190.
    See also Teratogen
Teratogen, 149, 158, 161, 166, 167, 189
Testes, 154. *See also* Reproduction
Tests. *See also specific methods*
    Ames, 90–91
    Draize, 86
    drunk driving, 65
    liver function, 116
    of organic produce, 177
Tetrachlorodibenzodioxin (TCDD),
    166–170
Thalidomide, 155, 158, 189–190, 214
Thanksgiving menu, 3
Therapeutic index, 34, 125, 187
Thiosulfate, 113
Thiourea, 145
Threshold
    carcinogens and, 135–136, 138,
        147–149
    concept of, 115–117, 119, 216
    defined, 117
    practical, 149–151
Thyroid cancer, 145, 146
Tight building syndrome, *see* Indoor air
    pollution
Time bomb, 124
Times Beach, 164
Tobacco, 7, 12, 75, 140, 151, 160, 183,
    207. *See also* Cigarettes
Tolerances
    to chemicals, 38, 41, 74, 78
    food (standards), 40
    immunologic, 75
Tooth, teeth, 49, 50, 145
Toxicant, defined, 31. *See also* Poisons
Toxicologist
    defined, 35
    employment, 44–45
Toxicology
    defined, 26–32
    empirical, 31
    history, 34–40
    origins of, 33
Transformers, 147, 169
Trans-science, 206
Tricarboxylic acid (TCA) cycle, 69

Trichlorophenol, 144, 167
Tri-ortho-cresyl phosphate (TOCP), 66
Tumors, 76, 89, 119, 131, 145, 146, 166,
    184. *See also* Cancer
    benign, 139
    classification of, 139
    diet and, 73
    liver, 170, 172, 215
Turpentine, 24, 153
Typhus, 174–175, 191

Ultraviolet rays, 133
United Auto Workers v. Johnson
    Controls Inc., 159
United States
    cancer in, 135, 141, 200
    chemicals bans and regulation, 24, 36,
        43, 70, 178, 179, 181
    cholera in, 200
    congenital malformations in, 51, 157,
        159
    drug approval, 95, 109, 141, 142–143,
        151, 188–189
    malaria in, 41
    organic food in, 42
    PCCs in, 172
    pesticides, 7, 14, 41, 175–177
    poison, 109–110
U.S. Department of Agriculture
    (USDA), 34, 36–38, 223
Uranium, 21, 22, 23
Urea, 4
Urinary bladder cancer, 145
Uterine cancer, 204

Vaccine
    anticancer, 141
    preventative, 35, 82, 141, 202–203
    therapeutic, 141
Vaginal cancer, 155
Vapors
    exposure to, 55–57
    formaldehyde, 182
    LC_{50} and, 85, 104
    mercury, 51, 61
Ventilation, 23, 134, 182–185, 227. *See
    also* Indoor air pollution
Vinegar, 20, 113

Printed in the United States of America
ED-03-29-12